# Mooring System Engineering for Offshore Structures

# Mooring System Engineering for Offshore Structures

Kai-Tung Ma

Yong Luo

Thomas Kwan

Yongyan Wu

ELSEVIER

Gulf Professional Publishing
An imprint of Elsevier

Gulf Professional Publishing is an imprint of Elsevier
50 Hampshire Street, 5th Floor, Cambridge, MA 02139, United States
The Boulevard, Langford Lane, Kidlington, Oxford, OX5 1GB, United Kingdom

**British Library Cataloguing-in-Publication Data**
A catalogue record for this book is available from the British Library

**Library of Congress Cataloging-in-Publication Data**
A catalog record for this book is available from the Library of Congress

ISBN: 978-0-12-818551-3

For Information on all Gulf Professional Publishing publications
visit our website at https://www.elsevier.com/books-and-journals

*Publisher:* Brian Romer
*Senior Acquisition Editor:* Katie Hammon
*Senior Editorial Project Manager:* Andrae Akeh
*Production Project Manager:* Anitha Sivaraj
*Cover Designer:* Miles Hitchen

Typeset by MPS Limited, Chennai, India

Working together
to grow libraries in
developing countries

www.elsevier.com • www.bookaid.org

# Contents

Biographies                                                              xv
Foreword                                                               xvii
Preface                                                                 xix
Acknowledgment                                                        xxiii

**1. Introduction**                                                       1

  **1.1 Overview**                                              1
  **1.2 History of offshore mooring**                           2
    1.2.1 Floating drilling—rapid growth in the 1960s and 1970s   2
    1.2.2 Floating production—deepwater boom in 2000s   5
    1.2.3 Technologies—enabling the migration to deeper water   5
    1.2.4 Industry standards—multiple codes needing
    harmonization                                      6
  **1.3 Floating drilling enabled by mooring**                   7
    1.3.1 Drilling semi                                 8
    1.3.2 Drillship                                     9
    1.3.3 Tender-assisted drilling (TAD)                9
  **1.4 Floating production enabled by mooring**                10
    1.4.1 Tension-leg platform (TLP)                   11
    1.4.2 Semisubmersible (semi)                       11
    1.4.3 Spar                                         12
    1.4.4 FPSO and FSO                                 13
    1.4.5 Catenary Anchor Leg Mooring (CALM) buoy      15
  **1.5 Differences between drilling and production**           16
  **1.6 Floating wind turbine**                                 17
  **1.7 Questions**                                             18
  **References**                                                18

**2. Types of mooring systems**                                          19

  **2.1 Overview**                                             19
    2.1.1 Temporary versus permanent moorings         20
    2.1.2 Catenary versus taut leg moorings           20
    2.1.3 Spread versus single-point moorings         21
  **2.2 Spread mooring system**                                 21
    2.2.1 Equally-spread versus clustered-spread moorings   23
  **2.3 Single-point mooring system**                          23
    2.3.1 Internal turret mooring system              24
    2.3.2 External turret mooring system              26

| | | |
|---|---|---|
| | 2.3.3 Disconnectable turret mooring system | 28 |
| **2.4** | **Other types of single-point mooring system** | **32** |
| | 2.4.1 Tower yoke mooring system | 32 |
| | 2.4.2 Catenary anchor leg mooring system | 34 |
| **2.5** | **Dynamic positioning and thruster-assisted systems** | **35** |
| | 2.5.1 Dynamic positioning system | 35 |
| | 2.5.2 Thruster-assisted mooring system | 38 |
| **2.6** | **Questions** | **38** |
| | **References** | **39** |

**3. Environmental loads and vessel motions**     **41**

| | | |
|---|---|---|
| **3.1** | **Loads on floating structures** | **41** |
| | 3.1.1 Mooring system to resist environmental loads | 41 |
| | 3.1.2 Site-specific environmental data | 43 |
| | 3.1.3 Loads in different frequency ranges | 43 |
| **3.2** | **Wind load** | **44** |
| | 3.2.1 Description of winds | 44 |
| | 3.2.2 Wind-induced forces | 46 |
| **3.3** | **Wave load and vessel motions** | **48** |
| | 3.3.1 Description of waves and swells | 48 |
| | 3.3.2 Wave-induced forces and motions | 51 |
| **3.4** | **Current load and vortex-induced motion** | **54** |
| | 3.4.1 Description of currents | 54 |
| | 3.4.2 Current-induced forces and vortex-induced motion | 56 |
| **3.5** | **Ice load** | **57** |
| | 3.5.1 Description of ices | 58 |
| | 3.5.2 Ice-induced forces and ice management | 58 |
| **3.6** | **Other topics on environment loads** | **59** |
| | 3.6.1 Directional combination of wind, waves, and current | 60 |
| | 3.6.2 Sensitivity study on wave period | 60 |
| | 3.6.3 Wave−current interaction | 61 |
| **3.7** | **Questions** | **61** |
| | **References** | **61** |

**4. Mooring design**     **63**

| | | |
|---|---|---|
| **4.1** | **Design basis** | **64** |
| | 4.1.1 Gather input data | 64 |
| **4.2** | **Design process** | **65** |
| | 4.2.1 Select mooring system type | 67 |
| | 4.2.2 Determine the profile (catenary or taut leg) | 67 |
| | 4.2.3 Design the mooring pattern | 68 |
| | 4.2.4 Design the mooring line composition | 71 |
| | 4.2.5 Optimize the mooring design | 73 |
| **4.3** | **Design considerations** | **74** |
| | 4.3.1 Limiting vessel offset | 75 |
| | 4.3.2 Minimizing line tension | 75 |
| | 4.3.3 Reducing fatigue damage accumulation | 76 |
| | 4.3.4 Avoiding clash or interference | 76 |

| | | | |
|---|---|---|---|
| **4.4** | **Design criteria** | | 77 |
| | 4.4.1 | Design codes | 77 |
| | 4.4.2 | Vessel offset requirement | 77 |
| | 4.4.3 | Strength design criteria | 78 |
| | 4.4.4 | Fatigue design criteria | 79 |
| | 4.4.5 | Operability requirement | 80 |
| **4.5** | **Engineering analysis and code check** | | 80 |
| | 4.5.1 | Mooring analysis load cases | 81 |
| **4.6** | **Questions** | | 82 |
| | **References** | | 83 |

| | | | |
|---|---|---|---|
| **5.** | **Mooring analysis** | | 85 |
| | **5.1** **Theoretical background** | | 86 |
| | | 5.1.1 Governing equations of mooring line | 86 |
| | | 5.1.2 Static solution—catenary equation | 87 |
| | | 5.1.3 Mooring line stiffness | 89 |
| | | 5.1.4 Mooring line dynamics | 91 |
| | | 5.1.5 Mooring system | 91 |
| | **5.2** **System modeling** | | 93 |
| | | 5.2.1 Modeling of floaters | 93 |
| | | 5.2.2 Modeling of mooring lines | 94 |
| | | 5.2.3 Modeling of risers | 96 |
| | | 5.2.4 Modeling of environments and seabed | 96 |
| | | 5.2.5 Analysis procedure | 96 |
| | **5.3** **Modeling of polyester rope stiffness** | | 97 |
| | | 5.3.1 Upper—lower bound model | 98 |
| | | 5.3.2 Static—dynamic model | 99 |
| | **5.4** **Quasistatic or dynamic analyses** | | 101 |
| | **5.5** **Strength analysis in frequency domain** | | 102 |
| | | 5.5.1 Response transfer functions | 102 |
| | | 5.5.2 Frequency-domain analysis procedures | 103 |
| | | 5.5.3 Limitation of frequency-domain analysis | 104 |
| | **5.6** **Strength analysis in time-domain** | | 105 |
| | | 5.6.1 Time-domain approach | 105 |
| | | 5.6.2 Analysis procedure | 106 |
| | | 5.6.3 Summary | 106 |
| | **5.7** **Uncoupled and coupled analyses** | | 107 |
| | | 5.7.1 Uncoupled analysis | 107 |
| | | 5.7.2 Coupled analysis | 108 |
| | | 5.7.3 Industry practice | 109 |
| | **5.8** **Response-based analysis** | | 109 |
| | **5.9** **Mooring software** | | 110 |
| | | 5.9.1 OrcaFlex by Orcina Ltd. | 110 |
| | | 5.9.2 DeepC/SESAM by DNV GL | 110 |
| | | 5.9.3 Ariane by Bureau Veritas | 111 |
| | | 5.9.4 Other tools | 111 |
| | **5.10** **Questions** | | 112 |
| | **References** | | 112 |

**6.    Fatigue analysis**                                                        115

   **6.1   Overview**                                                      115
          6.1.1   Miner's rule                                         117
   **6.2   Fatigue resistance of mooring components**                     117
          6.2.1   $T-N$ curves for chain, connectors and
                  wire ropes                                            118
          6.2.2   $S-N$ curves for chain and wire ropes                 119
          6.2.3   $T-N$ curve for polyester ropes                       120
          6.2.4   Comparison between $T-N$ and $S-N$ curves             121
   **6.3   Fatigue analysis in frequency domain**                         122
          6.3.1   Simple summation approach                            123
          6.3.2   Combined spectrum approach                           124
          6.3.3   Dual narrow band approach                            125
   **6.4   Fatigue analysis in time domain**                              125
   **6.5   Fatigue analysis procedure**                                   126
   **6.6   Vortex-induced motion fatigue**                                128
          6.6.1   Mechanism of vortex-induced motion                   128
          6.6.2   Vortex-induced motion fatigue assessment             129
   **6.7   Out-of-plane bending fatigue for chain**                       132
          6.7.1   Mechanism of out-of-plane bending fatigue            132
          6.7.2   Out-of-plane bending fatigue assessment              134
   **6.8   Questions**                                                    136
   **References**                                                        136

**7.    Model tests**                                                            139

   **7.1   Types of model tests**                                         140
          7.1.1   Ocean basin model test                              140
          7.1.2   Wind tunnel test                                    141
          7.1.3   Towing tank test                                    141
          7.1.4   Ice tank test                                       142
   **7.2   Principle of model test**                                      143
          7.2.1   Scale factor                                        144
   **7.3   Capability of model basin facilities**                         145
          7.3.1   Wind generation                                     145
          7.3.2   Wavemaker                                           145
          7.3.3   Current generation                                  146
   **7.4   Limitations of model test**                                    147
   **7.5   Mooring system truncation**                                    148
          7.5.1   Mooring truncation                                  148
          7.5.2   Truncation design                                   149
          7.5.3   Limitations due to truncation                       149
          7.5.4   Other truncation methods                            150
   **7.6   Hybrid test method**                                           150
          7.6.1   Hybrid method                                       150
          7.6.2   Basic principle                                     150
          7.6.3   Numerical tools                                     151

**7.7 Model test execution** 152
    7.7.1 Model preparation 152
    7.7.2 Environment calibration 152
    7.7.3 Data collection and processing 152
**7.8 Questions** 153
**References** 153

**8. Anchor selection** 155

**8.1 Overview** 155
    8.1.1 Available anchor types 155
    8.1.2 Anchor design considerations 157
    8.1.3 Soil characterization 157
**8.2 Suction piles** 158
    8.2.1 Holding capacity of suction piles 159
    8.2.2 Suction pile installation 160
**8.3 Driven piles** 161
    8.3.1 Holding capacity of driven piles 162
    8.3.2 Driven pile installation 163
**8.4 Drag embedment anchors** 163
    8.4.1 Advantages and limitations of drag embedment anchors 164
    8.4.2 Holding capacity of drag embedment anchors 165
    8.4.3 Drag embedment anchor installation and recovery 165
**8.5 Vertically loaded anchors** 166
    8.5.1 Vertically loaded anchor for permanent and temporary moorings 166
    8.5.2 Holding capacity of vertically loaded anchors 167
    8.5.3 Vertically loaded anchor installation 168
**8.6 Suction embedded plate anchors** 168
    8.6.1 Advantages and limitations of suction embedded plate anchor 168
    8.6.2 Suction embedded plate anchor installation 169
**8.7 Gravity installed anchors** 170
    8.7.1 Torpedo anchor 170
    8.7.2 OMNI-Max anchor 171
**8.8 Questions** 173
**References** 173

**9. Hardware—off-vessel components** 175

**9.1 Mooring line compositions** 175
**9.2 Chain** 176
    9.2.1 Studlink versus studless 177
    9.2.2 Chain grades 177
    9.2.3 Manufacturing process 178
**9.3 Wire rope** 180
    9.3.1 Six-strand versus spiral strand 181
    9.3.2 Corrosion protection 182
    9.3.3 Termination with sockets 183

| | | |
|---|---|---|
| **9.4** | **Polyester rope** | **183** |
| | 9.4.1   First use of polyester mooring in deepwater | 185 |
| | 9.4.2   Rope constructions | 185 |
| | 9.4.3   Polyester stretch | 187 |
| **9.5** | **Other synthetic ropes** | **187** |
| | 9.5.1   Nylon rope | 188 |
| | 9.5.2   High modulus polyethylene rope | 188 |
| | 9.5.3   Aramid rope | 190 |
| | 9.5.4   Considerations for moorings in ultradeep waters | 190 |
| **9.6** | **Connectors** | **191** |
| | 9.6.1   Connectors for permanent moorings | 191 |
| | 9.6.2   Connectors for temporary moorings | 193 |
| **9.7** | **Buoy** | **195** |
| **9.8** | **Clump weight** | **196** |
| **9.9** | **Questions** | **197** |
| | **References** | **197** |

| | | |
|---|---|---|
| **10.** | **Hardware—on-vessel equipment** | **199** |
| **10.1** | **Tensioning systems** | **199** |
| | 10.1.1   Fairlead and stopper | 201 |
| | 10.1.2   Hydraulic or electric power unit | 201 |
| | 10.1.3   Chain locker | 202 |
| **10.2** | **Chain jack** | **203** |
| **10.3** | **Chain windlass** | **204** |
| | 10.3.1   Movable windlass (or chain jack) | 204 |
| **10.4** | **Wire winch** | **206** |
| | 10.4.1   Drum winch | 206 |
| | 10.4.2   Traction winch | 207 |
| | 10.4.3   Linear winch | 209 |
| **10.5** | **In-line tensioner** | **209** |
| **10.6** | **Summary** | **212** |
| **10.7** | **Questions** | **212** |
| | **References** | **212** |

| | | |
|---|---|---|
| **11.** | **Installation** | **215** |
| **11.1** | **Site investigation** | **215** |
| | 11.1.1   Geophysical survey | 216 |
| | 11.1.2   Geotechnical survey | 216 |
| **11.2** | **Installation of permanent mooring** | **217** |
| | 11.2.1   Phase I—installation of pile anchors | 217 |
| | 11.2.2   Phase II—prelay of mooring lines on seabed | 219 |
| | 11.2.3   Phase III—hook-up of mooring lines to floating production unit | 222 |
| **11.3** | **Deployment and retrieval of temporary mooring** | **225** |
| | 11.3.1   Rig mooring system for mobile offshore drilling unit | 226 |
| | 11.3.2   Preset mooring system for mobile offshore drilling unit | 228 |

**11.4 Installation vessel** 229
  11.4.1 Anchor handling vessel 229
  11.4.2 Anchor handling vessel incident—capsizing
    of Bourbon Dolphin 230
**11.5 Questions** 231
**References** 232

## 12. Inspection and monitoring 233

**12.1 Inspection** 233
  12.1.1 Regulatory requirements 234
**12.2 Inspection schedule** 234
  12.2.1 As-built survey for permanent mooring 235
  12.2.2 Periodic surveys for permanent mooring 235
  12.2.3 Periodic surveys for Mobile Offshore Drilling
    Unit mooring 236
**12.3 Inspection methods** 236
  12.3.1 Difference between Mobile Offshore Drilling
    Unit and permanent moorings 236
  12.3.2 General visual inspection 237
  12.3.3 Close-up visual inspection 239
  12.3.4 Nondestructive examination techniques 240
  12.3.5 Advanced three-dimensional imaging 240
**12.4 Inspection of mooring components** 241
  12.4.1 Inspection of chain 241
  12.4.2 Inspection of wire rope 243
  12.4.3 Inspection of fiber rope 244
  12.4.4 Inspection of connecter and anchor 245
**12.5 Monitoring** 245
  12.5.1 Regulatory requirements 246
  12.5.2 What and how to monitor 246
**12.6 Monitoring methods** 247
  12.6.1 Method 1—monitoring visually 247
  12.6.2 Method 2—monitoring tension 248
  12.6.3 Method 3—monitoring vessel position 248
**12.7 Monitoring devices** 249
  12.7.1 Load cell 250
  12.7.2 Inclinometer 250
  12.7.3 Global Positioning System—based system 251
**12.8 Questions** 252
**References** 252

## 13. Mooring reliability 255

**13.1 Mooring failures around the world** 256
**13.2 Probability of failure for permanent moorings** 261
  13.2.1 Estimated $P_f$ for permanent moorings 262
  13.2.2 System versus component failures
    (multiline vs single-line breaks) 263

| | | |
|---|---|---:|
| **13.3** | **Failure spots for permanent moorings** | 264 |
| **13.4** | **Probability of failure for temporary moorings** | 265 |
| | 13.4.1   Estimated $P_f$ for mobile offshore drilling unit moorings | 266 |
| | 13.4.2   Improving mobile offshore drilling unit mooring reliability | 267 |
| **13.5** | **Failure spots for temporary moorings** | 269 |
| **13.6** | **Reliability of mooring components** | 270 |
| | 13.6.1   Percentage distribution of mooring failures by component type | 270 |
| | 13.6.2   Percentage distribution of chain failures by cause | 272 |
| **13.7** | **Wide variety of failure mechanisms** | 273 |
| | 13.7.1   Deficient chain from manufacturing | 274 |
| | 13.7.2   Chain with severe corrosion | 274 |
| | 13.7.3   Fatigued chain due to out-of-plane bending | 275 |
| | 13.7.4   Knotted chain due to twist | 275 |
| | 13.7.5   Chain damaged from handling | 276 |
| | 13.7.6   Operation issues | 276 |
| **13.8** | **Questions** | 277 |
| **References** | | 277 |

| | | |
|---|---|---:|
| **14.** | **Integrity management** | 281 |
| **14.1** | **Mooring integrity management** | 282 |
| | 14.1.1   Managing mooring performance | 282 |
| | 14.1.2   Assessing hazards and performing risk assessment | 283 |
| **14.2** | **Incident response** | 284 |
| | 14.2.1   Define response actions | 285 |
| | 14.2.2   Include a sparing plan | 286 |
| | 14.2.3   Predefine installation procedures and contracting plan | 287 |
| | 14.2.4   Include procedures for readiness check of equipment | 287 |
| **14.3** | **Life extension** | 287 |
| | 14.3.1   Life extension for a floating facility and its mooring system | 288 |
| | 14.3.2   Fitness assessment of mooring component | 289 |
| **14.4** | **Ways to improve mooring integrity** | 291 |
| | 14.4.1   Perform rigorous inspection and maintenance | 291 |
| | 14.4.2   Equip with monitoring system | 293 |
| | 14.4.3   Share lessons learned | 294 |
| | 14.4.4   Improve codes and standards | 294 |
| **14.5** | **Questions** | 295 |
| **References** | | 296 |

| | | |
|---|---|---:|
| **15.** | **Mooring for floating wind turbines** | 299 |
| **15.1** | **Concepts of floating offshore wind turbines** | 300 |
| | 15.1.1   History of concept development | 300 |
| | 15.1.2   Spar-buoy type | 301 |

|  |  |  |
|---|---|---|
| 15.1.3 | Semisubmersible type | 301 |
| 15.1.4 | Tension leg platform type | 302 |
| 15.1.5 | Comparison of concept types | 303 |
| **15.2** | **Mooring design** | 304 |
| 15.2.1 | Mooring type | 304 |
| 15.2.2 | Mooring line material | 305 |
| 15.2.3 | Anchor selection | 306 |
| **15.3** | **Mooring design criteria** | 307 |
| 15.3.1 | Design return period | 307 |
| 15.3.2 | Optional redundancy | 307 |
| 15.3.3 | Other requirements | 308 |
| **15.4** | **Mooring analysis** | 308 |
| 15.4.1 | Environmental forces and load cases | 308 |
| 15.4.2 | Aerodynamic loads | 310 |
| 15.4.3 | Time-domain mooring analysis | 311 |
| **15.5** | **Design considerations** | 312 |
| 15.5.1 | Fatigue | 312 |
| 15.5.2 | Corrosion | 313 |
| 15.5.3 | Installation | 313 |
| 15.5.4 | Tensioning | 313 |
| 15.5.5 | Overall project cost | 313 |
| **15.6** | **Questions** | 314 |
| **References** | | 314 |
| Index | | 317 |

# Biographies

**Kai-Tung Ma**

Dr. Kai-Tung (KT) Ma is a principal advisor at a major operator. He earned his BSc from National Taiwan University and his MSc and PhD from the University of California at Berkeley. He has around 25 years of experience and has previously worked for a few consulting firms as an engineer or manager, and as a senior engineer at a major classification society. He has published over 40 papers and 4 patents in the areas of mooring engineering, reliability of marine structures, and drilling riser design. KT is a fellow of SNAME and serves on the API and ISO mooring committees as chair and member, respectively. He is an adjunct professor at the National Taiwan University.

**Yong Luo**

Dr. Yong Luo is the founder and president of his own company which provides engineering services to the offshore industry. With over 30 years' experience, he previously worked for several large offshore companies. Dr. Luo has published over 70 papers and is a visiting professor at Shanghai Jiao Tong University and Harbin Engineering University in China. He earned his BSc degree from Shanghai Jiao Tong University, China, his MBA from the University of Leicester, United Kingdom, and his PhD from the University of Strathclyde, United Kingdom.

**Thomas Kwan**

Dr. Thomas Kwan has over 40 years' experience in the offshore industry. He worked for a major operator on mooring and riser systems. From 1982 to 2006 he served as chair for the API mooring committee, leading the development of API RP-2SK, API RP-2I, and ISO 19901-7. He has been a lecturer in the United States and Singapore teaching mooring system technology. He earned his BSc from Chu Hai College, Hong Kong, and his MSc and PhD from the University of Houston, United States. He has published many papers. Dr. Kwan received the Albert Nelson Marquis Lifetime Achievement Award in 2018.

**Yongyan Wu**

Dr. Yongyan Wu is a senior principal naval architect at a major engineering company in Houston as the mooring system lead. He holds a PhD from the University of Hawaii at Manoa, United States, and MSc and BSc degrees from Shanghai Jiao Tong University, China. He previously worked as an engineer at a major classification society. He has published over 20 papers in wave mechanics, wave/structure interaction, and stationkeeping systems and is a registered professional engineer in the State of Texas. Dr. Wu serves as a technical editing panelist on the mooring committee for API and the offshore technical committee for SNAME.

# Foreword

So many of us working in the offshore arena during the last four decades have constantly been challenged by the deceptive complexity of moorings. What to the casual observer looks like a simple structural line element, has, packed within it, every possibly complexity in concept (catenary, semi-taut, taut), material (chain, steel wire, fiber), and analysis (time domain, frequency domain, hybrid, extreme, fatigue, creep) to challenge the designer. From rules of thumb, we have moved to highly complex, precise and accurate analysis accounting for multiple physical processes in ultradeep waters reaching 3 km. In terms of equipment, the offshore industry has accelerated the move from simple mooring equipment from three millennia of shipping history to the most ingenious anchors and winches.

A book that encapsulated such a broad sweep of moorings has never been attempted until now. Some of the few people in our industry who could have pulled it off are the authors; I have had the pleasure and honor to have worked with most of them, and indeed the pride in supporting them, however small, in their march towards greater understanding. They represent the academic rigor necessary to be technically correct; while in their roles in some of our industry's most prestigious oil companies, consultants, contractors, and Classification Societies, they have had the great fortune to have been able to champion and realize their ideas.

The book comes at a critical juncture in the energy transition. In the last nearly three-quarter century of offshore hydrocarbon exploration and production, electricity from renewable energy sources has gone from nearly zero to 25% of world production. A rapidly increasing contributor to that is offshore wind. While offshore wind energy has made a major impact in Europe using fixed substructures, its global impact will only come with moored floating offshore wind in the Americas and Asia, in deeper waters. It is therefore gratifying to see that the authors have addressed this topic in a specific chapter in this book. This work will go towards bridging the knowledge transition between offshore oil and gas and offshore wind.

With elegantly direct writing, coupled with rich illustrations and clear tables, the book is a joy to read. It will be valuable as a text book, a design guide, a reference to go in and out of, and above all an authoritative document to keep within one's reach as a practitioner. I thank the authors for having brought so much of the knowledge of moorings into this one book, which will be of fantastic benefit not only for the oil and gas industry but also for the offshore wind industry.

**R.V. Ahilan**
*Joint CEO, LOC Group, London,*
*United Kingdom*

# Preface

This book deals with the design of mooring systems, from the viewpoint of a design engineer, and thus it covers what is required to properly address this task. It describes the systems and their performance features and it addresses the calculations that are needed for the design, going from the Classification Societies requirements to the computer programs that are available to conduct different types of calculations. It is thus an important reference work.

The book starts by introducing the history and background of offshore moorings. It then explains how mooring systems enable offshore operations such as floating drilling, floating production, and floating wind turbine power generation. Chapter 2 describes the types of mooring systems, and includes a comparative analysis of their performance and range of applicability.

Chapter 3 deals with the environmental loads and the vessels' motions. It discusses the wind and wave loads on the vessels (both first and second order) and their induced motions. The loads due to currents are discussed with regards to their effect on the vessels and on the moorings. Ice loads are also introduced and suggestions are made for the design of the moorings to avoid being affected by ice.

Having introduced the basics about the mooring systems and the loads acting on them, Chapter 4 deals with mooring design with the focus on moorings for mobile offshore units and permanent production facilities. It describes how to set up the design basis and then goes through the various steps of design providing the information necessary to accomplish them.

Chapter 5 deals with the analysis of mooring systems and is a chapter with more mathematical formulations for the mooring line dynamics and strength. It discusses frequency domain and time domain analysis as well as coupled and uncoupled analysis between mooring and floater. It finishes by introducing commercial software that is available to conduct those analyses.

Following mooring analysis, fatigue of the mooring lines is considered. The traditional approach of counting stress cycles associated with the application of Miner's rule is explained. Fatigue calculations are described in the frequency domain and in the time domain, associated with cycle counting methods. Vortex-induced fatigue is discussed, as well as out-of-plane bending of chains.

Model tests are normally an important part of the design process and they are discussed in Chapter 7. The typical set of tests is described, followed by a discussion of the characteristics of model basins and their influence on the interpretation of test results. The scale of the tests does not generally allow very deepwater moorings to be properly scaled and thus the established approach is the truncated formulation, which is discussed in the book. Hybrid methods, combining experimental results with numerical ones can be an alternative, as discussed in the rest of the chapter.

Anchor selection is the topic of Chapter 8, which starts with the description of the various anchors available and their main features. The details about the components of the various types of moorings are discussed in the next chapter, as well as their terminators and connections, and additional components such as clump weights and buoyancy floaters. Then the equipment existing on the vessels to deal with the moorings is described and their functions and limitations are discussed. This covers tensioning systems, chain jacks, chain windlass, wire winch, and in-line tensioners.

The installation procedures are described in Chapter 11, which starts with the site investigation, installation of permanent mooring, and deployment and retrieval of temporary mooring. This finishes up by describing anchor handling vessels, in particular their equipment and their role in the installation phases.

Inspection and monitoring are considered afterwards, including the requirements of Classification Societies, and the various inspection methods. Monitoring methods and devices are then described and their role and limitations are established.

Chapter 13 deals with the reliability of moorings, starting by discussing the record of mooring failures to derive the lessons learned and the probability of failure that has been present in the industry practice. Analysis is made for permanent and temporary moorings and the components more prone to failures are identified. The chapter concentrates on the analysis of service experience, not covering how those calculations could be made from first principles.

The next chapter discusses the integrity management of mooring systems connected to a "permanent" floating system used for the drilling, development, production, or storage of hydrocarbons and this is the last chapter dealing with this type of structure. This includes inspection, maintenance, monitoring, and repair, and the related topic of life extension is also contemplated.

The final chapter deals explicitly with floating wind energy platforms, which have their own specific features that require a different treatment of the moorings. The various types of platforms presently in use are described and the specific aspects of their mooring are described. The fact that they typically operate in water depths between 50 and 150 m has consequences on the type of mooring systems that are adopted. The methods of mooring analysis are described, which couple the aerodynamic forces on the turbine blades, with hydrodynamics of the platform and of the mooring.

Overall this is a very nice book written by industry authors and it is very comprehensive and detailed. One can find what is necessary in the various phases of the life of moorings, from the design to installation and to maintenance and life extension. The text covers what needs to be done and the tools that are required to do that. It is very useful in that it covers various detailed aspects that result from the authors experience and that will certainly be very useful to readers.

Each chapter has a list of references that is not very long but does contain the necessary background material, which comes predominantly from OTC Conference proceedings, which reinforces the industrial nature of the text.

The book is also very useful to students and academics as it contains information about the procedures that are being used in the industry, which is very often a good starting point for research studies.

**Carlos Guedes Soares**
*Distinguished Professor, Centre for Marine Technology and*
*Ocean Engineering (CENTEC), Instituto Superior Técnico,*
*Universidade de Lisboa, Lisboa, Portugal*

# Acknowledgment

The purpose of the book is for general education and training. To the knowledge of the authors, there is a lack of a comprehensive technical book that is dedicated to the subject of offshore mooring engineering. This book is written in response to such a need. It has been written for university students as a textbook and for practicing engineers as a reference book. Mooring engineering is a multi-discipline subject of naval architecture, ocean, civil, and mechanical engineering, and the subject matter of this book can serve as the basis of a graduate level course. Students should have attended the basic courses in stress analysis, hydrodynamics, mechanics of materials, and dynamics of structures as prerequisites to fully comprehend the material.

For the completion of the book, lots of credits should go to my three coauthors who shared equal burden with me in preparing the materials. Dr. Yong Luo is the founder of his own engineering service company, and is a visiting professor at Shanghai Jiao Tong University and Harbin Engineering University in China. Dr. Tom Kwan has been teaching short courses at University of Texas at Austin for many years. Dr. Yongyan Wu has been the lead mooring engineer designing various types of mooring systems at Aker Solutions in Houston.

Also, the contribution from our reviewers was immense. To facilitate the review of the manuscripts prepared by the four authors, a book club was formed with the following members: Paul Devlin, Menno van der Horst, Leopoldo Bello, Wei-ting Hsu, Wei-liang Chuang, Hongmei Yan, Devin Witt, Qinzheng Yang, and Ming Yang. Regular meetings were held to review the manuscripts, where the members debated and argued over every technical point. Outside of the book club, we received valuable comments or feedback from the following individuals: Cedric Brun, Prof. Yong Bai, Bob Gordon, Arun Duggal, Roger Basu, Haobing Guo, Zhengyong Zhong, Sam Ryu, Peter Leitch, and many others. Support and patience from the spouses of the authors allowed the time-consuming writing job to be completed smoothly; thanks to Sufen Chang, Li Zhu, Irene Kwan, and Haoshan (Sue) Guo.

Among all the reviewers, I would like to express my deepest gratitude to the lead reviewer, Paul, who came out the comfort of his retirement to attend the book club meetings and provided the in-depth review of the materials prepared by the authors.

**Kai-Tung Ma** (馬開東)
*Houston, December 2018*

# Chapter 1

# Introduction

## Chapter Outline

1.1 Overview 1
1.2 History of offshore mooring 2
   1.2.1 Floating drilling—rapid growth in the 1960s and 1970s 2
   1.2.2 Floating production—deepwater boom in 2000s 5
   1.2.3 Technologies—enabling the migration to deeper water 5
   1.2.4 Industry standards—multiple codes needing harmonization 6
1.3 Floating drilling enabled by mooring 7
   1.3.1 Drilling semi 8
1.3.2 Drillship 9
1.3.3 Tender-assisted drilling (TAD) 9
1.4 Floating production enabled by mooring 10
   1.4.1 Tension-leg platform (TLP) 11
   1.4.2 Semisubmersible (semi) 11
   1.4.3 Spar 12
   1.4.4 FPSO and FSO 13
   1.4.5 Catenary Anchor Leg Mooring (CALM) buoy 15
1.5 Differences between drilling and production 16
1.6 Floating wind turbine 17
1.7 Questions 18
References 18

## 1.1 Overview

The ocean provides us with valuable natural resources, such as crude oil and natural gas under the seabed. As many discoveries of large oil/gas fields are made offshore, floating structures built for exploration or production have become popular. Over the past decades, there has been a steady rise in the demand for floating platforms such as FPSOs (floating production storage and offloading), semisubmersibles, spars, and TLPs (tension-leg platform). A key element for these floating platforms is the mooring system, which is the subject covered by the present book.

The vital requirement for a mooring system is its ability to keep a floating structure on station under specific environmental conditions to allow various operations such as drilling, production, offloading, and wind power generation to be safely conducted. It is not an easy task for mooring engineers to design a system to meet such a requirement, because they constantly face challenges in areas of design, engineering, manufacturing, installation,

Mooring System Engineering for Offshore Structures. DOI: https://doi.org/10.1016/B978-0-12-818551-3.00001-6

operation, inspection, monitoring, maintenance, and repair [1]. These challenges will be thoroughly discussed chapter by chapter in this book.

This chapter introduces the history and background of offshore moorings. It then explains how mooring systems enable offshore operations such as floating drilling, floating production, and floating wind-turbine power generation.

## 1.2   History of offshore mooring

Mooring systems are made up of lines, connectors, tensioning equipment, and anchors. They have a long history of being used for station-keeping of floating vessels in various situations. For example, they have been used to berth boats or ships at quaysides, often referred to as quayside moorings. This book, however, focuses on "offshore moorings" for Mobile Offshore Drilling Units (MODUs), floating production units, and some other types of permanent floating structures. These offshore moorings have a relatively short history compared to, for instance, traditional ship related moorings.

### 1.2.1   Floating drilling—rapid growth in the 1960s and 1970s

The first MODU was the Mr. Charlie, which started drilling in 1954. It was a submersible barge built specifically to float on its lower hull for transportation to location. It had to run a sequence of flooding the hull stern down in order to rest on the bottom to begin drilling operations [2,3]. It was rated for 40-ft. water depth, but strictly speaking it was not a floating drilling system.

The very first "floating" drilling vessel to use subsea well control was the Western Explorer (Fig. 1.1) owned by Standard Oil of California (now Chevron), which spudded its first well in 1955 in the Santa Barbara Channel. The mooring lines are vaguely visible in front of the bow in the picture. The spudding from a floating vessel may be considered as the first milestone in the history of floating drilling.

With the Mr. Charlie (bottom-founded) and Western Explorer (floating) as the first MODUs, another concept for a MODU showed up in the form of a "jackup." This type of unit floats to location on a hull with multiple legs sticking out under the hull. Once on location, the legs are electrically or hydraulically jacked down to the ocean bottom, and then the hull is jacked up out of the water. With this approach, a stable platform is available to drill from. The Gus I, as an example, was the first jackup built for drilling in 1956 for 80-ft. water depth. In the 1990s, "premium" or "enhanced" jack-ups were designed and built, which could drill in deeper depths—greater than 400 ft. of depth.

Shell Oil saw the need to have a more "motion-free" floating drilling platform in the deeper and stormier waters of the Gulf of Mexico (GOM). They noticed that submersibles like the Mr. Charlie had much smaller motions compared with those of monohulls. They had the idea of putting a mooring system on a submersible, thus converting it to a semisubmersible

**FIGURE 1.1**  First moored drillship with subsea well control—Western Explorer. *Courtesy of the Alden J. Laborde Family.*

(or semi) that floats. Thus, in 1961, the submersible Bluewater I was converted to a semi. Then came the Ocean Driller, the first semi built from the keel up. The Ocean Driller went to work for Texaco (now Chevron) in 1963. The unit was designed for approximately 300 ft. of water depth, with the model tests of the hull done in a swimming pool.

In the 1960s, owners of deepwater drilling barges and self-propelled drillships employed mooring systems consisting of chain and wire rope connected to six or eight anchors. Using anchors became a common practice for station-keeping for those semis and drillships [3]. During the decade, these mooring systems allowed floating units to drill in much deeper water than their bottom-supported counterparts, that is, jack-ups. The increased water depth imposed challenges on mooring arrangements. Longer mooring lines became heavier and more difficult to handle. Large vessel offset due to weather could overstress the riser due to deflection and possibly lead to failure. Mooring design and analysis need to be performed to allow drilling risers to stay connected to the blowout preventer (BOP) on the seafloor. Station-keeping became an important engineering discipline, and offshore mooring was born as an integral part of the offshore industry.

Semisubmersibles and other types of drilling vessels went through an evolution of designs. Most of the first-generation units could sit on the seafloor or drill from the floating position. The shape and size of the first semis varied widely, as designers strived to optimize vessel motion characteristics, rig layout, and other considerations. In the early 1970s, second-generation semis were designed and built with newer, more sophisticated mooring and subsea equipment. The design generally was for 600-ft. water depth or deeper. In the mid-1980s, a number of third-generation semis were designed and built that could moor and operate in greater than 3000 ft. of water depth and more severe environments. The displacement of these units went from approximately 18,000 long tons in the 1970s to more than 40,000 tons in the 1980s. In the late 1990s, the fifth-generation units, became even larger at a displacement greater than 50,000 long tons. Fig. 1.2 shows a drilling semi with full dynamic positioning (DP) and moored configurations. The unit can operate in extremely harsh environments.

**FIGURE 1.2** Fifth-generation drilling semisubmersible designed for deepwater exploration. *Courtesy of Transocean.*

## 1.2.2   Floating production—deepwater boom in 2000s

Oil and gas have been produced from offshore locations since the late 1940s. Bottom-founded structures such as fixed platforms and compliant towers were initially used, which are limited to water depth of about 1000−1800 ft. As exploration and production moved to deeper waters and more distant locations, four types of floating production systems (FPS) came to play including the following.

- *TLP*—The concept of tension leg platform (TLP) was first applied in early 1980s. It is a vertically moored floating structure normally used for water depths greater than 300 m (about 1000 ft.) and less than 1600 m (about 5200 ft.). The first TLP was the Hutton TLP installed in North Sea in 1984.
- *Semi*—The first semisubmersible floating production platform was the Argyll FPS converted from the Transworld 58 drilling semisubmersible in 1975 for the Hamilton Brothers Argyll oil field in the North Sea. The first purpose-built semisubmersible production platform was for the Balmoral field in the North Sea in 1986.
- *Spar*—It is a type of floating production platform with a cylindrical shape and deep draft, which makes it less affected by wind and waves and allows for both production through deck mounted wellheads (dry tree) and subsea (wet tree) production. The first spar was the Neptune spar installed in 1997 in GOM.
- *FPSO*—A floating production storage and offloading (FPSO) unit is a floating vessel used by the offshore oil and gas industry for the production and processing of hydrocarbons, and for the storage and offloading of oil. The first oil FPSO was the Shell Castellon, built in Spain in 1977. Today, over 270 vessels are deployed worldwide as FPSOs.

All four types of floating production systems, TLP, semisubmersible, spar, and FPSO, have experienced significant growth over the years since the first installation. In particular, FPSOs have a much larger number of units installed in the world than the other three types. Floating production moved into deep water at 6000−8000 ft. of depth during the year 2000s. Ironically, the year 2000s was also the decade when several major mooring failures surprised the industry, as they were caused by different novel failure mechanisms. Those failures are reviewed and discussed in later chapters of this book relating to reliability and integrity.

## 1.2.3   Technologies—enabling the migration to deeper water

Initially floating drilling was conducted in shallow water of less than 100 ft., and then gradually moved to slightly deeper waters. Up to the 1980s, drilling vessels were moored in water depths no more than a few hundred feet.

During that period, wire ropes and chain were the components used on every mooring job. Technologies in mooring and station-keeping have improved significantly since then.

Now, in deeper waters, drilling can be conducted by vessels employing DP systems, that use computer-controlled, motor-driven propellers, called "thrusters," to counter the wind, waves, and current loads. They respond automatically to signals coordinated with acoustic beacons placed on the seafloor. Note, however, early DP systems were not very reliable. DP drilling vessels often experienced malfunction of the system such as drift-off due to power blackout or drive-off due to system (or human) errors. DP vessels are now getting more reliable, and they can drill in waters as deep as 12,000 ft.

On the other hand, moored drilling and production vessels can also operate in deeper water with the advances of mooring technologies. One significant breakthrough for deepwater mooring is the technological advancement of synthetic fiber rope mooring. Polyester and other fiber ropes were studied for deepwater moorings in the early 1990s. The studies showed that polyester rope has desirable weight and stiffness characteristics for use as mooring lines. The first use of polyester ropes in a permanent mooring system was attempted successfully by Petrobras in the mid-1990s. Temporary moorings for MODUs also began to use polyester lines in the GOM in the early 2000s. The first permanent applications of polyester mooring systems in the GOM were the Mad Dog and Red Hawk production platforms, installed in 2004. Today, polyester mooring has become the most commonly used mooring system for deepwater floating production around the world.

In addition to those polyester moorings, there are other technology developments that have enabled the industry's migration to deep water over the last 50 years. For example, mooring chain has been advanced from ORQ to R5 grades with much higher break strength. Anchors have been improved as well. In the early days of floating drilling, small conventional drag anchors were used exclusively. Today, high-efficiency drag anchors and Vertically Loaded Anchors are available for drilling operations. For floating production units, more powerful anchors that can take high vertical loads such as suction piles are widely deployed. All these technological advancements enabled the offshore industry to venture into deeper waters and harsher environments.

### 1.2.4  Industry standards—multiple codes needing harmonization

The first mooring standard, Recommended Practice 2P (API RP 2P), was published in 1984 by the American Petroleum Institute (API). It was developed to address design and analysis of spread mooring systems for floating drilling platforms. Around the same time, the industry began to use floating production platforms for oil production instead of the conventional fixed jackets, depending on water depth. API RP 2FP1 was published in 1987 to address the stricter requirements needed in the mooring design for floating production platforms. Later, the first edition of API RP 2SK was published in 1995 to combine API RP 2P and API RP 2FP1 into one document, which also provided additional guidance based on the technological advancement at the time. The publishing of RP 2SK was a milestone in codes and standards for offshore mooring.

The third edition of API RP 2SK [4] was published in 2005 with a significant improvement over the previous editions. It became arguably the most widely used mooring design code in the industry. It provided guidance for the first time on several important issues such as mooring hardware requirements, chain corrosion allowance, clearance criteria, anchor design and installation, design for vortex-induced motions, global performance analysis guidelines, and MODU mooring criteria for GOM hurricane season. RP 2SK was used as the basis for some other codes, such as ISO 19901-7 published by the International Standards Organization (ISO).

There were several other mooring standards developed by class societies, such as the American Bureau of Shipping, Det Norske Veritas, and Bureau Veritas. To a certain degree, the multiple standards caused some confusion for designers, as they could have different requirements on safety factors, metocean criteria, or corrosion allowance. To resolve the issue of conflicting mooring standards, API and ISO initiated an effort to develop unified standards. As part of this effort, a panel was created in 1995 to develop an ISO mooring standard, and the first edition of ISO 19901-7 was published in 2005 as the result. ISO later published the second edition in 2013 [5]. API has a separate standard, RP 2I, which is still the most widely used standard for in-service inspection of mooring hardware [6], and ISO 19901-7 does not include in-service inspection.

Harmonization of multiple codes and standards has never been an easy task [7]. It is even more challenging for mooring design standards, as offshore mooring is a relatively young engineering discipline that is still constantly evolving. Significant joint industry effort may be required to develop up-to-date and consistent mooring standards to meet the industry's needs [1,7,8].

## 1.3  Floating drilling enabled by mooring

Offshore drilling operations are conducted by MODUs, which can be categorized into at least four types: jack-up barge, drilling semi, drillship, and tender-assisted-drilling (TAD) vessel. Jack-up barges use their vertical legs to stand on the seabed and raise their hulls above the sea surface, so they may not be considered as one of the vessel types for floating drilling. The other three types use either a mooring system, a DP system, or a thruster-assisted mooring system to provide station-keeping for floating drilling. The operation to drill an exploratory well is normally of short duration, lasting 30–90 days. Therefore there is a need for the vessel to move periodically from one well site to another. The frequent moves require the mooring system to be designed for easy retrieval and redeployment. The relocation of MODUs is typically done by towing, unless the vessel is equipped with a DP system that allows self-propelling.

### 1.3.1 Drilling semi

Most floating drilling vessels are of semisubmersible types which are designed with good stability and seakeeping characteristics. A semisubmersible retains most of its buoyancy from ballasted, watertight pontoons located below the ocean surface and wave action. Vertical columns connect the pontoons and operations deck. With its hull structure submerged at a deep draft, the semisubmersible is less affected by wave loadings than a drillship. With a small water-plane area, however, the semisubmersible is sensitive to load changes, and therefore must be carefully designed to maintain stability.

The drilling operation is conducted through a drilling riser, which is connected to a BOP for well control. The drilling riser can have an angle mainly due to the horizontal offset of the floating vessel, and the angle must be controlled to stay within a limit [9], as shown in Fig. 1.3. The function of the mooring system is to ensure that the vessel is kept within these offset limits

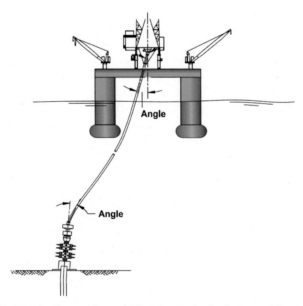

**FIGURE 1.3** Vessel offset causing a drilling riser to bend. *Courtesy of Stress Engineering Services, Inc.*

so that the drill pipe can be rotated inside the riser without damaging the upper and lower flex joint of the riser.

In recent years, drilling semis used in deep water (say, 2000−8000 ft.) and ultradeep water (say, 8000 ft. and deeper) have become very large, and some of them are equipped with highly-advanced DP systems. However, most of them still have a mooring system available onboard. The mooring system allows the semis to drill in shallow water areas, where DP systems may not be able to maintain the riser angle within an allowable limit.

## 1.3.2   Drillship

Drillships, which can be spread moored or dynamically positioned, have a long history of being used in floating drilling operations. Because of the relatively poor motion characteristics of a ship-shaped vessel, moored drillships are seldom used today.

In recent years, drillships used in deep water and ultradeep water have become very large, and have no mooring system. They are equipped with highly-advanced DP systems instead. These DP systems maintain the position of a drillship within a small specified tolerance by controlling their thrusters to counter the wind, wave, and current forces. Without a mooring system, drillships may not be able to service shallow water areas due to the angle limit on drilling risers. Discussion of DP systems is beyond the scope of this book.

## 1.3.3   Tender-assisted drilling (TAD)

The TAD concept is used to reduce the requirements on deck space and dead weight for a fixed platform, TLP, semi, or spar. It allows reductions in deck load and space requirements since the drilling package is not permanently installed on the production platform. A TAD vessel brings a drilling package together with the consumables such as drilling fluid (mud). The vessel is moored alongside the production platform to provide the equipment and support, as shown in Fig. 1.4.

Early tenders were barge shaped. Semisubmersible hulls are used more now because they offer better station-keeping and vessel motions compared with barge-shaped hulls. TADs are seeing new uses on deepwater floating production platforms, such as spars and TLPs. In addition to supporting drilling and completion operations, tenders can also provide living quarters for the offshore operation personnel.

The design of the tender mooring requires special attention because of the close proximity between the tender and the host platform. An asymmetric mooring pattern may have to be used to avoid mooring lines clashing with the host platform, and the vessel offsets need to be carefully controlled to

**FIGURE 1.4** Moored tender vessel (right) providing drilling support to a production and drilling platform (left). *Courtesy of Atlantica Tender Drilling Ltd.*

avoid exceeding the limits of the connections between the two platforms such as gangway bridges, mud hoses, and hawser lines.

## 1.4 Floating production enabled by mooring

Fixed platforms were initially used to produce oil and gas in water depth up to about 1200 ft. The concept of compliant tower (Fig. 1.5) was developed for production in water depths beyond 1200 ft. However, the water depth capability of compliant tower is still limited to about 1800 ft. As exploration moved to deeper waters, floating production systems were deployed.

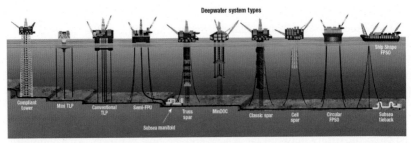

**FIGURE 1.5** Alternative production facilities from fixed platform to floating systems. *Courtesy of Wood.*

Mooring systems are used to keep floating production platforms on station. There are several types of floating production platforms, including TLP, semi, spar, FPSO, and offloading buoy. They are introduced in the following sections.

### 1.4.1   Tension-leg platform (TLP)

The concept of TLP is a vertically moored floating platform that is suitable for water depths between 1000 and 5000 ft. (Fig. 1.5). The platform is permanently moored by tendons grouped at each corner of the hull. The tendons, which are tubular steel pipes, have relatively high axial stiffness such that most vertical motion of the platform is eliminated. This allows the platform to have the production wellheads on deck (dry trees), connected directly to the subsea wells by rigid risers. This allows for a simpler well completion, and gives better control over the production along with easier access for downhole interventions.

The tendons provide the function of mooring lines, but are more like top-tensioned riser pipes. Their installation requires precision and careful handling. In comparison, conventional mooring components such as chain can be easily connected with piles and prelaid on seabed long before the scheduled hook-up time.

### 1.4.2   Semisubmersible (semi)

A semisubmersible platform is also known as semisubmersible, semisub, or semi. It is a specialized vessel designed for offshore drilling, oil production, heavy lifting, accommodation, or a combination of these functions. When oil wells are drilled and completed by drilling vessels, production semis are towed to the field and hooked up to their permanent mooring systems. Sometimes, production semis are intentionally designed with a built-in drilling rig so that development drilling can continue on the same platform after production has started (first oil). Semisubmersibles are stable and cost-effective platforms. As the offshore oil and gas development moved into deeper water, the use of semisubmersible platforms became increasingly popular because of their spacious deck area to accommodate large topside equipment and the ease of topside—hull integration at quayside.

Besides the conventional design with a shallow draft, there is an improved version called deep-draft semi. The latter was developed to further reduce the vertical motion of the platform. The reduced motion helps to improve the performance of steel catenary risers, and thus can be more cost-effective, and may even allow for a dry tree solution. Fig. 1.6 shows a deep-draft semi.

**FIGURE 1.6** Deep-draft semisubmersible with its mooring lines and fairleads clearly shown. *Courtesy of Aker Solutions.*

### 1.4.3 Spar

Spar is a type of floating production platform typically used in deep waters. The deep draft design of spars makes them less affected by waves. The good vertical motion characteristics allow for both dry tree and subsea (wet tree) production. A classic spar as shown in Fig. 1.7 consists of a large-diameter, single vertical cylinder supporting a high deck. The cylinder is ballasted at the bottom by a chamber filled with a material that is denser than water to lower the center of gravity of the platform, thus improving stability. Additionally, the spar hull is encircled by helical strakes to mitigate the effects of vortex-induced motion caused by current. Spars are permanently anchored to the seabed by way of a spread mooring system composed of either a chain−wire−chain or chain−polyester−chain configuration.

There are two other types of spar: truss spar, and cell spar. A truss spar has a shorter cylindrical "hard tank" than a classic spar and has a truss structure connected to the bottom of the hard tank. This truss structure consists of four large orthogonal "leg" members with X-braces between each of the legs and heave plates at intermediate depths to provide damping of vertical motions. At the bottom of the truss structure, there is a relatively small keel, or soft tank, that houses the heavy ballasting material. The majority of spars

**FIGURE 1.7**   Classic spar.

are of the truss type. A third type of spar, the cell spar has a large central cylinder surrounded by smaller cylinders of alternating lengths. At the bottom of the longer cylinders is the soft tank housing the heavy ballasting material, similar to a truss spar. The cell spar is intended for small field production.

### 1.4.4   FPSO and FSO

A floating production storage offloading (FPSO) system is a ship-shaped vessel used for producing hydrocarbons and storing crude oil. An FPSO vessel is designed to receive hydrocarbons produced from its wells, nearby platforms, or subsea equipment. It has equipment on deck to process the hydrocarbons. It also stores the oil until it can be offloaded onto a shuttle tanker or, less frequently, transported through an export pipeline. FPSOs are often the preferred concept for frontier regions where there is no local pipeline infrastructure to export oil. A vessel used only to store oil without processing capability is referred to as a floating, storage, and offloading (FSO).

The development of the FPSO system has generally gone through five phases, namely:

- From 1976 to 1985, for the early stages of development, it became possible to deploy FPSOs in all areas for offshore oil and gas production through the single-point mooring system;
- From 1986 to 1994, during the FPSO growth period, the FPSO mooring technology was rapidly developed at this stage, associated with the increase of FPSO units by an average of two or more per year;
- From 1995 to 1998, during the expansion period of FPSOs, the number of FPSOs in this phase increased significantly, increasing at an average rate of more than eight units per year;
- From 1999 to the present, the number of FPSOs has increased rapidly, and technological breakthroughs have been achieved. The operating water depth increased from the initial 100 m to 3000 m. The current record of FPSO operating water depth is Shell's Stone FPSO in the US GOM in a water depth of 2920 m;
- Especially after 2002, the FPSO concept has been extended to other types of operations, such as in the form of LPGFPSO, LNGFPSO, Floating Storage and Regasification Unit, Floating Production Drilling Storage and Offloading system.

It can be seen from the above history that the FPSO system has been continuously developed and widely adopted [11]. The station-keeping system for FPSOs can be a spread mooring, a turret mooring, or a disconnectable turret mooring. The third type can be disconnected from the vessel before the arrival of severe weather (such as hurricane).

Produced oil can be transported to the mainland either by shuttle tanker or export pipeline. When a shuttle tanker is chosen to transport the oil, the offloading of oil from FPSO to the shuttle tanker can be carried out by using an offloading buoy. Alternatively, oil can be transferred from the FPSO to the tanker directly through tandem or side-by-side arrangements. The tandem offloading arrangement (as shown in Fig. 1.8) is often the preferred option due to its better safety than the side-by-side arrangement. The latter has to deal with the issue of close proximity, where two floating vessels are positioned right next to each other. In the opinions of some mooring engineers, tandem offloading may be the only suitable option for turret moored FPSOs. However, for Floating Liquefied Natural Gas, which is a barge-shaped floating facility to convert offshore natural gas into liquefied natural gas (LNG), the side-by-side offloading arrangement is often used.

**FIGURE 1.8** Turret moored floating production storage and offloading with a shuttle tanker connected in tandem. *Courtesy of Bluewater.*

### 1.4.5    Catenary Anchor Leg Mooring (CALM) buoy

A Catenary Anchor Leg Mooring (CALM) buoy system consists of a buoy that is kept on station by a number of catenary mooring lines anchored to the seafloor. The mooring lines are typically all chain in shallow water, but can be a combination of chain, wire rope, and fiber rope in deep water locations. Riser systems or flow lines from the seafloor are attached to the underside of the buoy. These systems use a hawser, typically a synthetic rope, to connect the buoy to the visiting tanker. Since the response of the CALM buoy is totally different than that of the tanker under the influence of waves, this system is limited in its ability to withstand environmental conditions. When sea states reach a certain magnitude, it is necessary to disconnect and cast off the tanker.

Since the early days of the offshore industry, the CALM buoy has been a successful system for the importing and exporting of oil. Initially, these buoys were moored in relatively shallow water at near-shore locations, often in harsh wave conditions. In recent years, the use of CALM buoys for offloading crude oil from FPSOs in deep water has become more and more common. The CALM buoy systems may be the most popular and widely used type of offshore loading terminal, with more than 500 installed to date. More details are introduced in Chapter 2, Types of mooring systems.

## 1.5 Differences between drilling and production

In terms of mooring designs, there are significant differences between mooring systems used for floating drilling and those for floating production. The major differences are in parameters such as design environment, floating vessel type, mooring component size, anchor choice, deployment frequency, and inspection method, as shown in Table 1.1.

These differences are mainly caused by the mobility requirement for drilling vessels. The nature of a MODU mooring system is that it should be easily moved from location to location. That implies that the MODU moorings

**TABLE 1.1** Differences in mooring design parameters between floating drilling and production.

|  | Floating drilling | Floating production |
|---|---|---|
| Design environment | 10-year storm (or 5-year in open water) | 100-year storm |
| Vessel type | Semi, drillship, barge | FPSO, semi, spar, TLP |
| Mooring chain size | Smaller, up to 4-in. diameter | Larger, up to 7-in. diameter |
| Anchor choice | Smaller, drag anchors | Larger, typically suction or driven piles |
| Deployment frequency | Deployed frequently for typical drilling duration of 1−3 months | Installed permanently for the design life |
| Inspection method | Hands-on during deployment or retrieval | Visual by underwater ROV |

*FPSO,* Floating production storage and offloading; *ROV,* remotely operated vehicle; *TLP,* tension-leg platform.

can't be overly robust due to weight. They have planned short service duration ranging from a few weeks or a few months to typically no more than 12 months [10]. Because of the short service duration, mooring systems for drilling operation are designed for a substantially lower environment of typically 10-year return period, whereas permanent moorings for production are typically designed for 100-year return period. The difference in design criteria may explain why drilling vessels have experienced a higher mooring failure rate due to environmental overloading than production platforms.

## 1.6    Floating wind turbine

Floating wind turbines are floating systems permanently stationed in open water to harvest wind energy in the form of electricity. Higher winds are available offshore compared to on land, so offshore wind power electricity generation is higher per amount of capacity installed. In addition, offshore wind power generation offers the advantages of more stable output and less noise and visual pollution. Unlike the typical usage of the term "offshore" in the marine industry, offshore wind power includes onshore water areas such as lakes, fjords and sheltered coastal areas, utilizing traditional fixed-bottom wind turbine technologies, as well as deeper-water areas utilizing floating wind turbines.

Initially fixed-bottom structures were used for offshore wind power generation. For locations with depths over about 60−80 m, fixed-bottom structures are uneconomical, and floating wind turbines with mooring systems are preferred (as shown in Fig. 1.9).

**FIGURE 1.9**   Floating wind turbines stationed by mooring systems. From Left to Right: Spar, Semi-submersible, and TLP. *Illustration by Josh Bauer, National Renewable Energy Laboratory.*

Hywind is the world's first full-scale floating wind turbine, installed in the North Sea off Norway in 2009. The hull design is a spar shape. Hywind Scotland, commissioned in October 2017, is the first operational floating wind turbine with a capacity of 30 MW. Other kinds of floating wind turbines have been deployed, and more projects are planned.

## 1.7 Questions

1. What are the purposes of a mooring system for floating drilling, floating production, and floating wind turbine?
2. In your opinion, would you call TLP tendons mooring lines? What are the reasons that support your opinion?
3. What are the main differences in mooring designs among floating drilling, floating production, and floating wind turbine?
4. How would you choose between DP and mooring? In the future, do you think that DP will replace moorings?
5. Can you think of other applications of offshore moorings that are not introduced in this chapter? For example, permanent mooring for weather buoys.

## References

[1] K. Ma, A. Duggal, P. Smedley, D. LHostis, H. Shu, A historical review on integrity issues of permanent mooring systems, in: OTC 24025, OTC Conference, May 2013.

[2] H. Veldman, G. Lagers, 50 Years Offshore, Published by PennWell Books, 1997.

[3] F. Jay Schempf, Pioneering Offshore: The Early Years, Published by the Offshore Energy Center and PennWell Custom Publishing, 2007.

[4] API RP 2SK, Recommended Practice for Design and Analysis of Stationkeeping Systems for Floating Structures, third ed., American Petroleum Institute (API), 2005.

[5] International Standard, Stationkeeping Systems for Floating Offshore Structures and Mobile Offshore Units, ISO 19901-7, second ed., 2013.

[6] API RP 2I, In-Service Inspection of Mooring Hardware for Floating Structures, third ed., 2008.

[7] H. Shu, A. Yao, K. Ma, W. Ma, J. Miller, API RP 2SK fourth edition — an updated stationkeeping standard for the global offshore environment, in: OTC 29024, OTC Conference, May 2018.

[8] C. Kwan, Mooring design standards — the past, present, and future, in: Proceedings of the 20th Offshore Symposium, Texas Section of SNAME, February 2015, Houston, Texas.

[9] API RP 16Q, Design, Selection, Operation, and Maintenance of Marine Drilling Riser Systems, second ed., 2017.

[10] K. Ma, R. Garrity, K. Longridge, H. Shu, A. Yao, T. Kwan, Improving reliability of MODU mooring systems through better design standards and practices, in: OTC 27697, Offshore Technology Conference, May 2017.

[11] J. Paik, A. Thayamballi, Ship-Shaped Offshore Installations: Design, Building, and Operation., Cambridge University Press, 2007.

# Chapter 2

# Types of mooring systems

## Chapter Outline

**2.1 Overview**    **19**
    2.1.1 Temporary versus
      permanent moorings    20
    2.1.2 Catenary versus taut leg
      moorings    20
    2.1.3 Spread versus single-point
      moorings    21
**2.2 Spread mooring system**    **21**
    2.2.1 Equally-spread versus
      clustered-spread moorings    23
**2.3 Single-point mooring system**    **23**
    2.3.1 Internal turret mooring
      system    24
    2.3.2 External turret mooring
      system    26

    2.3.3 Disconnectable turret
      mooring system    28
**2.4 Other types of single-point
mooring system**    **32**
    2.4.1 Tower yoke mooring
      system    32
    2.4.2 Catenary anchor leg
      mooring system    34
**2.5 Dynamic positioning and
thruster-assisted systems**    **35**
    2.5.1 Dynamic positioning
      system    35
    2.5.2 Thruster-assisted
      mooring system    38
**2.6 Questions**    **38**
**References**    **39**

## 2.1 Overview

Offshore moorings are an essential part of the station-keeping systems developed for the exploration and production of offshore oil and gas resources. Apart from restraining the floating structure at its designated location, a mooring system also limits the floater's excursion to ensure the integrity and operability of drilling and production facilities such as production risers, drilling risers, and umbilicals. In addition to balancing the environmental loads, the restoring force generated by the mooring system also counterbalances operational loads such as those required during pipe laying. Offshore mooring systems can be designed for a wide range of conditions from a harsh environment, such as the North Sea, to a mild environment, like the Gulf of Thailand or Offshore West Africa. They can also be designed for a wide range of water depths from a few meters to over 3000 m.

Mooring System Engineering for Offshore Structures. DOI: https://doi.org/10.1016/B978-0-12-818551-3.00002-8

### 2.1.1 Temporary versus permanent moorings

Based on the duration of the offshore operation, mooring systems can be divided into the following two categories:

1. Temporary mooring system—Suitable for drilling semis, drill ships, pipe laying vessels, crane vessels, flotels, logistics supply vessels, etc. with its station-keeping duration from a few days to several months.
2. Permanent mooring system—Suitable for a variety of long-term floating structures. Depending on the field design life, it may be necessary to maintain station-keeping at a host location for several years to several decades.

### 2.1.2 Catenary versus taut leg moorings

Depending on the profiles and configurations, mooring systems can be grouped into *catenary mooring systems* and *taut leg mooring systems* (Fig. 2.1). As shown in the figures below, the catenary mooring system has a

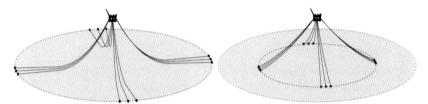

**FIGURE 2.1** (Left) Catenary mooring system. (Right) Taut leg mooring system.

line profile with part of the mooring line lying on the seabed in the static equilibrium position. Due to the self-weight of the mooring line, the mooring leg forms a catenary shape which generates the necessary compliance to cope with floater's static offset and dynamic motions. The catenary mooring system is the most widely used system, especially in shallow to medium-depth waters.

The taut leg mooring system has no line lying on the seabed in the static equilibrium position, and the mooring lines are taut from the anchor at seabed to the fairlead on the floater. Therefore the anchor footprint is smaller, and the mooring system uses less line material compared to the catenary mooring system. However, as the lines are taut, the compliance to floater offset and dynamic response is mostly from the line tensile stretch. Therefore a taut leg system in shallow water may be too stiff and can increase the line tension excessively. It is more suitable for deep or ultradeep water applications.

### 2.1.3    Spread versus single-point moorings

Based on the mooring system's requirement to restrict the floater's heading, mooring systems can be divided into *spread mooring systems* and *single-point mooring (SPM) systems* (as shown in Fig. 2.2). A spread mooring

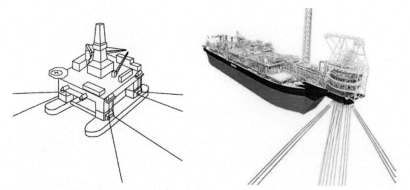

FIGURE 2.2    (Left) Typical spread mooring system. (Right) Typical single-point mooring system. *Courtesy of Vryhof and SOFEC.*

system has multiple mooring lines connecting the floater to the seabed, restricting the floater's position as well as its heading. A SPM system has one or multiple mooring lines connecting the floater's center of rotation to the seabed and thus allows the floater to weathervane about this center of rotation to head into the prevailing environment to minimize environmental loading.

## 2.2    Spread mooring system

The spread mooring is a natural extension of the traditional mooring of ships. Multiple mooring lines are distributed around the floating structure, restricting the floater's offset and heading to ensure its designated operation. In designing the layout of the spread mooring system, the preferred heading is determined by the local environmental conditions.

Relatively speaking, a spread mooring system is simple and economical, and does not require complicated rotational mechanical systems. Once the anchors are deployed, the position and direction of the floating vessel are effectively restricted, and risers and umbilical systems can be installed and operated.

Most of the Mobile Offshore Drilling Units and some floating production systems use the spread mooring system for station-keeping purposes. For example:

1. A drilling semisubmersible typically uses 8 or 12 mooring lines in four groups connected from the four columns to the seabed. They are often referred to as $4 \times 2$ (four sets of two legs as shown in Fig. 2.2) and $4 \times 3$.

2. A drillship typically uses four groups of mooring lines connected from the bow and stern of the vessel to the seabed.
3. A production semisubmersible or floating production storage and offloading (FPSO) typically uses 12 or 16 mooring lines in four groups connected from the four columns to the seabed. They are often referred to as 4 × 3 (as shown in Fig. 2.3) and 4 × 4.

**FIGURE 2.3** Clustered-spread mooring in 4 × 3 (4 clusters with 3 lines in each cluster) pattern for an floating production storage and offloading. *Courtesy of SBM Offshore.*

4. A production spar typically uses 9 or 12 mooring lines in three groups connected from the outer shell of the platform to the seabed. They are often referred to as 3 × 3 and 3 × 4.

For ship-shaped floating vessels with spread moorings, the mooring lines are connected at the bow and stern and spread outward, which can restrain the lateral offset and heading of the hull. In theory, a spread mooring system can be used in any geographic area as long as it has sufficient strength; in reality, if the ship-shaped floating structure is subject to large lateral environmental loads (i.e., in a beam sea condition), the mooring lines may not be able to withstand the excessive loading. Therefore although the spread mooring system is cost-effective, its application to large ship-shaped floating structures is only suitable for areas of a benign environment or where the directionality of wind, wave, and current is persistent, such as in offshore West Africa.

In addition, the spread moored FPSO has the main disadvantage of lower availability for offloading operations than the turret-moored FPSO [1]. When the direction of the environment is not favorable, shuttle tankers may have difficulty approaching and staying connected to the FPSO.

To overcome this problem, it is typical to have a CALM buoy next to the spread moored FPSO in the field setup for export operations of a shuttle tanker.

### 2.2.1  Equally-spread versus clustered-spread moorings

The selection of mooring line spread has to consider a number of factors such as design, installation, the directionality of environment, mooring material cost, etc. The equal spread design in which all mooring lines are spread symmetrically with a uniform spread angle is for ease of design and possibly installation. These are referred to as *equally-spread mooring* systems. On the other hand, the mooring lines can often be grouped into three or four groups to improve the mooring performance and to create large angular space for risers and other subsea facilities. These are often referred to as *clustered-spread mooring* systems.

## 2.3  Single-point mooring system

The application of the SPM system to the FPSO began in the 1970s. The vigorous development of the FPSO and its mooring system was within the proceeding decade. By the year 2018, there were around 150 FPSOs (not counting FSOs) deployed in various parts of the world, including the North Sea, Offshore West Africa, Offshore Brazil, the South China Sea, and the Gulf of Thailand. Most of them use SPM systems for station-keeping. The weathervaning capability of the SPM system reduces the loads due to wind, waves, and current, so that the dimensions of the mooring lines are effectively minimized.

SPM systems for FPSOs have great adaptability to work in different environmental conditions. However, they are technically challenging and expensive to build. Currently, companies that possess the state-of-the-art SPM system technology include SBM Offshore, SOFEC, Bluewater, and NOV APL.

The SPM system is well suited for crude offloading as well. A shuttle tanker can be connected with an FPSO either in tandem or side-by-side mode with the tandem offloading being more popular and safer. The shuttle tanker tied to an FPSO (or FSO) can rotate freely around the SPM, and the offloading operation is convenient, safe, efficient, and reliable.

Typically, the SPM system has the following two functions:

1. Station-keeping—The FPSO is kept at a location within limited offset from the offshore oil field operation site by the mooring restoring force.
2. Transfer of liquid and power—Through the special liquid swivel, electric slip ring, etc., the produced liquid, liquid injection to wells, electrical power supply, and communication signals can be continuously

transmitted from the seabed to the FPSO or from the FPSO to the wells while the FPSO weathervanes.

There are many different types of SPM systems. Based on different working characteristics and the location of the turret, the turret system can usually be divided into two main types: *internal turret system* and *external turret system*.

A turret is a steel structure with an upper part directly connected to the FPSO topsides, a mid-part connected to the FPSO hull via bearings, a lower part connected to the mooring lines, and decks supporting a stack of swivels. The turret-moored FPSO can rotate around the inner bearing of the turret in reaction to wind, waves and current action. Flexible risers are connected to the rigid tubes at the bottom of the chain table by flanges.

### 2.3.1 Internal turret mooring system

Among the mooring systems for FPSOs, the internal turret system is most commonly used in harsh environments with a large number of risers and umbilicals. The internal turret is generally located near the bow of the FPSO. The turret sits inside the moonpool of the FPSO hull, and the mooring lines are connected at the bottom of the turret via a chain table.

The internal turret can be supported by a single bearing or double bearings. The main components of the FPSO internal turret system include:

1. *Turret and its bearings*—The turret has the lower part that connects the mooring lines, and the upper part that is attached to the FPSO hull inside the moonpool. The FPSO hull and the turret are connected by the upper main bearing and the lower bearing, allowing the FPSO to freely revolve around the turret to head into the prevailing environment. Depending on the requirements of the riser and umbilical systems, the turret diameter is usually between 5 and 20 m.

2. *Chain stoppers*—The chain stopper connects the top end of the mooring line to the bottom of the turret structure (chain table). The top segment of the mooring line, typically chain, passes through the chain hawse and terminates at the chain stopper.

3. *Swivel (fluid transfer) system*—The swivel system is the centerpiece of the entire turret system, located in the upper part of the turret. The produced crude, the water and gas injections, signals to control subsea wells, power supply to subsea systems, etc. are all transmitted through a swivel system (also known as FTS).

4. *Gantry structure*—A structural platform that sits on the FPSO deck and surrounds the top of the turret to accommodate equipment such as swivel drivers, connecting pipes, winches for tightening mooring lines, power control facilities, etc.

Schematic diagrams of the internal turret system are shown in Figs. 2.4–2.6.

The advantage of the internal turret system is that the turret diameter can be larger than that of an external turret system, providing plenty of space for a large number of risers. The typical limit on the number of risers may be

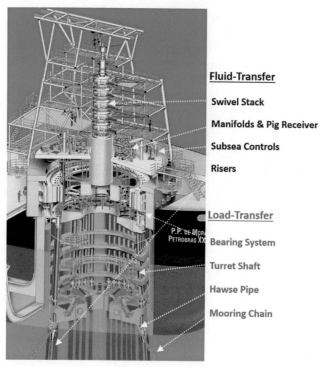

**FIGURE 2.4**  Internal turret mooring system in a floating production storage and offloading. *Courtesy of SOFEC.*

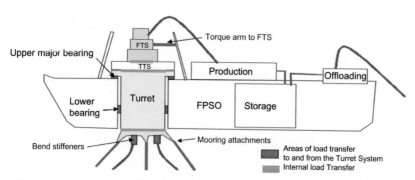

**FIGURE 2.5**  Illustration of FPSO and turret mooring system.

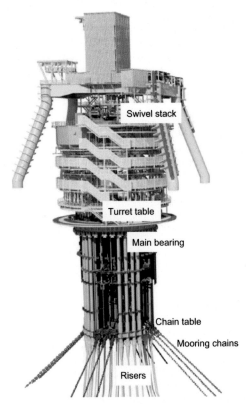

**FIGURE 2.6** Swivel stack, bearing, and chain table in a turret mooring system. *Courtesy of SBM Offshore.*

about 100, whereas the limit for an external turret system could be about 20 [1]. The inner turret is well protected by the hull. As the position of the internal turret is closer to the center of the FPSO, the vertical movement of the mooring connecting position is less than that of an external turret, thus reducing the loads on mooring lines and risers.

The disadvantage of the internal turret system is that the presence of the turret reduces the tank capacity. Another disadvantage is that the turret has an impact on the hull structure, especially in the case of a large turret diameter where the hull moonpool would need special reinforcement.

### 2.3.2 External turret mooring system

The external turret system (Fig. 2.7) is similar to the internal turret system except that the turret is located outside the FPSO hull.

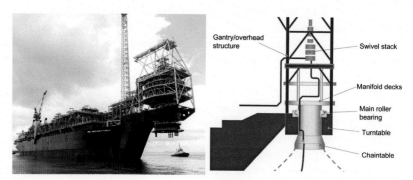

**FIGURE 2.7**    External turret system. *Courtesy of SOFEC and SBM Offshore.*

Based on the different positions of the turret extension, the external turret system can be further divided into two types: *raised external turret mooring* (Fig. 2.7) and *submerged external turret mooring* (Fig. 2.8).

**FIGURE 2.8**    Submerged external turret system. *Courtesy of SBM Offshore.*

The raised external turret is the most common type that is directly connected to the overhanging structure at the bow or stern of an FPSO well above the waterline. In order to prevent the mooring lines from colliding with the hull, the overhanging structure should be long enough. Alternatively, part of the FPSO bow needs to be cut off. The advantage of the raised turret is that under normal working conditions, the bearings, chain stoppers, and flexible riser connections are located above water for easy inspection and maintenance. The downside is that if the turret extension structure is too long the upper part of the flexible riser would be greatly affected by extreme waves.

Since the external turret is positioned well above the waterline, it is frequently used in shallow waters, say below 50 m, to increase the effective water depth.

The submerged external turret mooring usually adopts a double bearing structure. One bearing is above the water, and the other is submerged. The chain stoppers are submerged to avoid potential collision between the mooring lines and the hull. The advantage is a small overhanging structure and flexible risers below the water surface, resulting in no concern of wave slamming. However, the requirement of a dual bearing system increases the complexity and the underwater mooring and riser connection increases the difficulty for installation, operation, and maintenance.

Due to its relatively small size, the external turret system can accommodate only a limited number of risers at about 20, whereas the internal turret system may hold up to about 100 [1]. However, an external turret system has several advantages over internal turret systems including the following:

1. It allows stand-alone fabrication and installation along the quayside, while the internal turret system can only be installed in the dry dock.
2. It does not affect the storage capacity of the hull.
3. For the raised external turret, the mooring and riser connections are above water, which eases inspection and maintenance.

Catering to different design requirements and metocean conditions, other types of SPM systems have also been developed, which are covered later in this chapter.

### 2.3.3 Disconnectable turret mooring system

Turret systems can also be divided into permanent turret systems and disconnectable turret systems (Fig. 2.9). For the permanent turret system, the

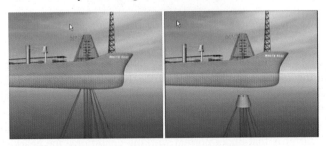

**FIGURE 2.9** Disconnectable turret system. *Courtesy of SBM Offshore.*

FPSO is permanently connected with its moorings and risers even in all extreme conditions. For the disconnectable turret system, the FPSO can be disconnected with its moorings and risers and sail away before the arrival of extreme conditions such as hurricane or iceberg conditions. This type of turret system inherently offers added safety but requires a complex mechanical system for disconnection.

Compared with the permanent system, the disconnectable turret system has the following features:

1. *Added safety due to the disconnect capability*—The disconnectable turret system is considered safer because the FPSO can effectively avoid extreme wind and waves by releasing the mooring lines and the risers, and sail away to safe harbor. After disconnection, the mooring lines and the risers sink to 50–100 m below the surface, and are less affected by the waves.
2. *Higher cost due to mechanical complexity*—The disconnectable system is more complicated and hence more expensive than the permanent as the disconnectable valves and connectors are costly. In addition, the entire installation process is more complicated, and more supporting facilities are needed.
3. *Complicated procedures for installation or disconnection operation*—The operation of the disconnectable system is highly complicated. It is necessary to avoid damage to any component during the disconnection and reconnection. The mooring lines and the risers are connected to the lower part of the buoy and it is difficult to adjust their lengths. Therefore the disconnectable mooring system does limit the use of fiber mooring lines that experience creep over the life of the mooring.

There are many designs of disconnectable turrets, and the common ones are:

*External Disconnectable Turret (RTM)*—The external disconnectable turret system (Fig. 2.10) is also called riser turret mooring (RTM) system. It is mainly composed of an overhanging turret, a joint head connector, a cylindrical riser buoy, mooring lines, a universal joint, a swivel joint, and a mechanical connecting device. It is suitable for sites subject to high frequency of tropical storms, thus requiring regular disconnection. The riser buoy may remain free-floating at the sea surface when disconnected and it must resist potentially harsh environmental conditions. The advantages of the system include low investment, easy manufacturing and demolition, and easy disconnection and reconnection. The disadvantage is that the reconnection is significantly affected by the vertical motion of the riser buoy. For reconnection, a cable is pulled up by a winch arrangement mounted at the bow.

**FIGURE 2.10**  Riser turret mooring system. *Courtesy of SBM Offshore.*

*Internal Disconnectable Turret*—The *BTM* (buoy turret mooring) system is one of the internal disconnectable turret systems, as shown in Fig. 2.11.

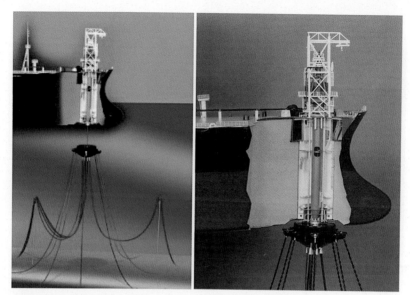

**FIGURE 2.11**   Buoy turret mooring system. *Courtesy of SBM Offshore.*

It consists of an inner turret embedded in the FPSO bow and a "spider" mooring buoy, which is connected to the turret by a structural connector. The spider buoy when disconnected provides sufficient buoyancy to support the weight of the mooring lines and risers. When the disconnection is activated, the spider buoy sinks below the water surface by the weight of the moorings and risers until the vertical equilibrium is reached. At the time of reconnection, the buoy is picked up at the bow of the FPSO and pulled through the turret using a winch. The advantage of this type of system is that the reconnection process is less affected by the waves since the operation is carried out underneath the vessel. The disadvantage is that the spider body is exposed to high hydrostatic pressure, and if mounted below the vessel keel, the hydrodynamic force acting on the extruded portion of the buoy may be a design concern.

The *STL* (submerged turret loading) or *STP* (submerged turret production) is another type of internal disconnectable turret system, as shown in Fig. 2.12. It consists of an underwater buoy that is conically shaped with

**FIGURE 2.12** Submerged turret production system. *Courtesy of NOV APL.*

integrated bearings. When the buoy is connected, it is embedded in the conical opening at the bottom of the hull to complete the coupled connection. When released, the buoy sinks to 30−50 m water depth. With the bearing integrated into the buoy, as opposed to the bearings integrated into the hull in the BTM design, the turret integration in the vessel is simplified. Due to its compact layout, the system is suitable for shallow or deep waters, and for working environments that require a rapid release.

When connected, the internal disconnectable turret system is identical to that of the permanent turret system. When disconnected, the mooring lines and risers are released from the FPSO. It should be noted that the risers need to be disconnected before the disconnection of buoy and turret. The disconnectable buoy, the mooring system, and the risers of the disconnectable turret system require careful design. Their interaction needs to be accurately modeled such that the buoy depth is well controlled, and the moorings and risers can cope with both the connected and disconnected conditions.

## 2.4 Other types of single-point mooring system

### 2.4.1 Tower yoke mooring system

The tower yoke mooring system uses a rigid jacket structure fixed on the seabed as the anchor point, and the tanker and the jacket are connected by a group of permanently jointed steel arm structures. The system has been developed for ultra shallow water depth (below 50 m) where a conventional catenary mooring system could not make the flexible risers work due to the limited water depth. (Note flexible risers need to maintain a certain predefined shape to avoid getting overstressed.) The main components of a soft yoke mooring system include:

1. Tower—A fixed jacket connected to the seafloor, the upper part of which is a turntable connected to the hull. The two parts are connected by bearings, allowing the turntable to rotate, which in turn allows the moored FPSO to freely rotate (weathervane) to the direction where the environmental force is minimal.
2. Mooring parts—Jointed structural members that connect the bearing on the tower to the bow or stern of the FPSO.
3. Production transmission system—Like the conventional turret system, the produced crude, the water and gas injections, signal to control subsea wells, power supply to subsea systems, etc. are all transmitted through a swivel system.

The advantage of the tower yoke mooring system is that it can accommodate many fixed steel risers (no flexible riser is required), the construction of the flow valves is relatively simple, and the system is relatively easy to install. The disadvantage is that the cost of the system increases rapidly with increasing water depth, so from a cost-effective point of view, this system is only suitable for shallow water applications.

The fork-type tower yoke mooring system is the most commonly used tower yoke mooring system (Fig. 2.13) and can be applied to FPSO or FSO. The fork structure includes a large ballast tank filled with water or concrete to provide the necessary restoring force to reduce hull offset.

**FIGURE 2.13**    Tower yoke mooring system. *Courtesy of Bluewater.*

## 2.4.2   Catenary anchor leg mooring system

With the growing crude-oil trade, petroleum products are transported around the world. For coastal areas where a deepwater port is not readily available, the application of catenary anchor leg mooring (CALM) System, as shown in Fig. 2.14, is often used to load and offload the crude-oil products. The

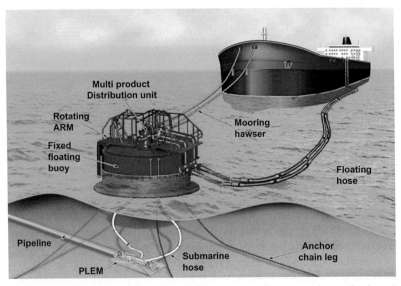

**FIGURE 2.14**   Catenary anchor leg mooring buoy temporarily connected to a shuttle tanker with hawsers. *Courtesy of SBM Offshore.*

CALM system can be designed to receive tankers of any size, including very large crude-oil carriers. Its main components include:

1. *Tanker hawser equipment* (soft mooring)—The tanker is temporarily moored to the CALM by hawsers during loading and offloading operations. The hawser mooring line is usually composed of nylon ropes and chain.
2. *Turntable and the buoy*—Allows the moored tanker to weathervane and carry out loading and offloading operations under various wind, wave, and current conditions without interruption. Although, the tanker will have to be disconnected if the seas get too rough.
3. *Floating hose*—One or more floating hoses connect to the tanker and the single-point piping system. According to Oil Companies International Marine Forum (OCIMF) [2], hose sizes (inner diameter)

typically range from 6 in. to a maximum of 24 in. *Underwater hose (riser)*—It is connected from a PLEM (Pipeline End Manifold) to a single-point piping system. The hose is designed to have sufficient length to accommodate the tidal range and CALM buoy motion. Through subsea pipelines, the PLEM connects to land terminals for oil and gas transmission.

4. *Anchor Leg Mooring System*—It may be considered as a permanent mooring system because its legs do not get retrieved or redeployed, even though a shuttle tanker comes to hook up only once every few days. It usually consists of mooring chains along with high holding-power drag anchors or pile anchors.

The CALM system has typically been used in waters that are close to the coast, shallow in water depth, and harsh in environmental conditions. The working water depth is usually 20−100 m. Since around the year 2000, the CALM mooring system has been extended to deep waters in excess of 1000 m off the coast of West Africa as an independent offloading system for spread moored FPSOs. Because of the large water depth, some CALM buoys use polyesters ropes rather than all chain to reduce the weight of mooring lines.

The advantages of the CALM system over conventional fixed terminals can be summarized as follows:

1. There is no need for a deepwater port, and there is no need for fixed dock facilities to provide loading and offloading operations.
2. The system can accommodate different tanker sizes, requires less human intervention, and has strong adaptability to the environment.
3. The system has a lower cost, as it can be installed in open sea areas that do not require protection.
4. It can transmit a variety of oil and gas products at the same time.
5. The system is easy to install and easy to operate. Only a small vessel is required to service the loading operation.

## 2.5  Dynamic positioning and thruster-assisted systems

### 2.5.1  Dynamic positioning system

One challenge in the design of station-keeping systems for ultradeep waters is the selection of the most cost-efficient system for the specified operational requirements. In addition to the spread moorings and SPM introduced above, a full dynamic positioning (DP) system is also an option for station-keeping (Fig. 2.15). DP systems have been extensively used by drilling

**FIGURE 2.15** Dynamic Positioning System for an FPSO, with thrusters attached to the hull. *Courtesy of SBM Offshore.*

units, that is, drill ships or drilling semis. Because the cost for traditional mooring systems becomes higher as water depth increases, a DP system can become a cost-effective solution, particularly for relatively short-term operations.

For production operations in deepwater, the capital cost of a mooring system including its installation can increase dramatically with an increase in water depth. In addition, seafloor congestion, poor geotechnical conditions, or short field life may result in the traditional mooring system not being the optimal solution. Thus, beyond certain water depths and for certain other conditions and applications, a full DP system may become more cost-effective than mooring systems [3]. However, if a production operation is long-term (i.e., permanent), constant powering of a DP system adds tremendous fuel costs, as DP is an active system that is constantly running. Therefore DP may be a viable option only for short-term operation such as an early production facility or an extended well-test facility, rather than a production operation for a full field development [3,4].

There have been several successful uses of DP systems for production operations. One example is Seillian DP FPSO in the North Sea. After 8 years of normal operation, the operator relocated it to the Roncador field in the Brazilian waters as an early production system. Another example is the use of DP FPSO "Munin" to temporarily resume the production of an oil field in the South China Sea after the moored FPSO "Nan Hai Fa Xian" was damaged by a typhoon in 2009. The field was brought back into production after 5.5 months of interrupted production. The DP FPSO system operated for more than 18 months and proved safe and effective [5]. At the time, it was a world record duration for an FPSO vessel operating in DP mode.

The DP system is out of the scope of this book, so only its operating principle is briefly described herein. A DP system maintains the position of a floating structure by means of thrusters. It applies precision instruments to measure the changes in displacement and azimuth of floating structures due to wind, wave, and current. The automatic feedback system processes and calculates the position feedback information, and controls the thrust and torque generated by several propellers. The floating structure returns to the specified position and the most favorable direction.

The main components of the DP system include:

1. Power steering system—Provides the driving force required for positioning.
2. Propeller system—By controlling the propulsion in transverse, longitudinal, and torsional directions, the floating structure is kept at the specified position. There are four typical types of propeller systems: azimuth thruster, jet propeller, cycloidal propeller, and fixed propeller.
3. Position measurement system—Track the position of the floating structure and feedback the position differential (real position vs target position) to the control system in real time.
4. DP control system—Positioning the floating structure to its design target by counteracting the environmental loads. The DP control system includes: a sensor system and a position reference system, such as measuring the wind speed and the offset of the floating structure. The collected data are processed, the discordant data discarded, and the high-frequency portion of the position signal is filtered out to determine the force acting on the floating structure, thereby transferring the force requirement to the propulsion system to counterbalance the external force.

The principle of the DP system is as follows:

1. The wind measurement system measures the wind speed and wind direction, and predicts the wind feedforward force/moment and feeds it into the DP control system.
2. The position measurement system tracks the floating structure's real-time position versus the target reference position to predict the required feedback force/moment to bring the position deviation to zero.
3. Superimposing the feedforward force/moment and the feedback force/moment, the thrust system distributes the thrust/moment to individual thrusters based on the principle of minimum power consumption.

**4.** The relevant sensor will feedback the working status of the wind, floater position, and thruster to the control system on a continuous basis.

## 2.5.2 Thruster-assisted mooring system

Some floaters are equipped with both DP and passive mooring systems. Either system can be used individually or jointly with an integrated positioning management system. Often the floater is equipped with manually operated or automated thrusters, and the output of the thruster forces can be utilized to assist the passive moorings. The thrusters-assisted mooring system consists of mooring lines and thrusters. The thrusters contribute to directly offset the environmental load and/or control the floater's heading, thus reducing mooring line forces and the floater's offset.

To determine the amount of thrust that can be allowed in the mooring calculations, several factors are to be considered, including the thruster's efficiency, the worst-case failure scenario, and the long-term thruster availability. The design standard defines the mean load reduction that can be used for spread moorings and for turret-moored floaters. For manual control of thrusters, the allowable thrust is further reduced to 70% on the basis that a manual system uses thrust in a fixed direction which will not necessarily be the most effective direction. For an automatic thruster system, 100% of the thruster output can be used to reduce the external force.

In the case of turret-moored ship-shaped floaters, maintaining a heading into the prevailing environment is the best way to reduce the mooring line loads.

## 2.6 Questions

1. Explain the difference between the catenary and taut leg mooring systems.
2. You are a naval architect designing a semisubmersible that will be used for drilling operations. A drilling derrick and a moonpool are obviously needed at the center of the deck layout. During drilling operations, a marine riser needs to stay right on top of the well. Do you choose a spread mooring system, a SPM system, or other? Why?
3. What are the advantages and disadvantages of internal and external turret mooring systems? Give one example when you would choose an internal turret system over an external one.
4. Draw a picture to show how crude oil travels from a riser, through a turret, all the way to a shuttle tanker.
5. What is the purpose of a CALM buoy system? What are the main components? Do you consider its anchor legs as a permanent or a temporary mooring system?

# References

[1]  J. Paik, A. Thayamballi, Ship-Shaped Offshore Installations: Design, Building, and Operation, Cambridge University Press, 2007.

[2]  OCIMF, Recommendations for Equipment Employed in the Bow Mooring of Conventional Tankers at Single Point Mooring, fourth ed., The Oil Companies International Marine Forum (OCIMF), 2007.

[3]  J. Lopez-Cortijo, A. Duggal, R. van Dijk, S. Matos, DP FPSO – a fully dynamically positioned FPSO for ultra deep waters, in: International Offshore and Polar Engineering Conference, ISOPE, Honolulu, 2003.

[4]  L. Poldervaart, B. Cann, J.H. Westhuis, A DP-FPSO as a first-stage field development unit for deepwater prospects I relative mild environments, in: Offshore Technology Conference, OTC 16484, Houston, May 2004.

[5]  J. Mao, Fast production recovery of a typhoon-damaged oil field in the South China Sea, in: CNOOC Limited, SPE Paper 172999-PA, vol. 3, no. (5), 2014.

# Chapter 3

# Environmental loads and vessel motions

## Chapter Outline

3.1 **Loads on floating structures** 41
  3.1.1 Mooring system to resist environmental loads 41
  3.1.2 Site-specific environmental data 43
  3.1.3 Loads in different frequency ranges 43
3.2 **Wind load** 44
  3.2.1 Description of winds 44
  3.2.2 Wind-induced forces 46
3.3 **Wave load and vessel motions** 48
  3.3.1 Description of waves and swells 48
  3.3.2 Wave-induced forces and motions 51
3.4 **Current load and vortex-induced motion** 54

3.4.1 Description of currents 54
3.4.2 Current-induced forces and vortex-induced motion 56
3.5 **Ice load** **57**
  3.5.1 Description of ices 58
  3.5.2 Ice-induced forces and ice management 58
3.6 **Other topics on environment loads** **59**
  3.6.1 Directional combination of wind, waves, and current 60
  3.6.2 Sensitivity study on wave period 60
  3.6.3 Wave−current interaction 61
3.7 **Questions** **61**
**References** **61**

A floating structure in open water experiences environment loads due to the existence of wind, waves, current, ice, etc. as shown in Fig. 3.1. Wind, waves, and current, as discussed in this chapter, are the most important environment parameters that generate loads on the floating structure. The mooring system is designed to provide the station-keeping capability to withstand these loads. Floating structures operating in ice-prone regions may be subjected to ice loading, which is introduced as well.

## 3.1 Loads on floating structures

### 3.1.1 Mooring system to resist environmental loads

The environmental loads acting on the floating system are ultimately resisted by the mooring restoring force. The magnitude and directionality

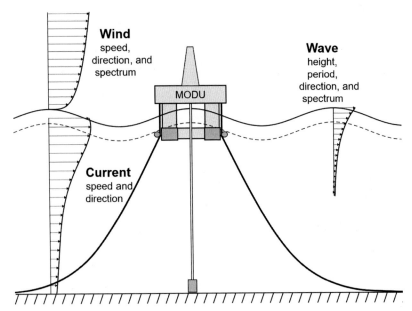

**FIGURE 3.1** Wind, wave, and current acting on an MODU. *MODU*, Mobile offshore drilling unit.

of the wind, waves, and current are the most important parameters for the mooring design. They affect the selection of mooring system type (e.g., spread or single-point mooring systems), and the overall mooring configuration.

A mooring system is designed to withstand the environmental loads so that the strength and fatigue requirements are met by all mooring components. In the strength design, a mooring strength analysis is performed to predict the mooring line tension, vessel offset, and anchor load under an extreme design environment of a specific return period. A return period, also known as a recurrence interval, is an estimate of the likelihood of an event to occur, such as a storm environment, an earthquake, or ice event. A storm of 100-year return period is expected to occur on average once every 100 years. The maximum calculated mooring line tension, vessel offset, and anchor load are then checked against the factors of safety specified by design standard.

In the fatigue design, a mooring fatigue analysis is performed under a set of environmental states that adequately represents the long-term statistics of the local environment to yield fatigue lives for the mooring components, which are then checked against factors of safety specified by the design standard.

### 3.1.2    Site-specific environmental data

The interaction of environmental phenomena, such as wind, waves, current, and ice, is site-specific. For strength analysis, the joint probability distribution describing these environments is considered in the development of the maximum design conditions (e.g., 100-year return period event). Of particular importance are the relationships among wind, wave, and current, and their relative directionalities. The mooring system should be evaluated for a suitable number of load cases which include the most unfavorable combinations of wind, wave, and current directions, consistent with the site-specific metocean characteristics. Note that there are regions governed by special environmental phenomena that may not be well represented by parameters with typical return period statistics. For example, some regions can be subject to sudden wind storms such as squalls, and other areas can be subject to occasional currents of high velocities. In these cases, the special occurrences are considered in determining the relevant maximum design conditions.

With the site-specific environmental data defined, the vessel offset and mooring line strength are evaluated during the mooring design. For fatigue analyses, sufficient wave data are collected to provide a scatter diagram that adequately represents the long-term statistics of the local environment. Other site-specific data required include water depth, soil properties, seafloor conditions, and marine growth profiles [1].

### 3.1.3    Loads in different frequency ranges

Environmental loads acting on floating structures can be categorized as follows according to their distinct frequency bands:

- *Steady loads* such as mean wind, current, and mean wave drift forces are constant in magnitude and direction for the duration of interest. Steady loads push the floating structure to an offset that is counterbalanced by the mooring restoring force.
- *Wave frequency cyclic loads* with typical periods ranging from 5 to 30 seconds. The loads result in wave frequency motions of the floating vessel, which create cyclic tensions that contribute to maximum mooring line tensions and fatigue damage accumulation in the mooring lines. In some situations, the moorings and risers provide additional damping to vessel motions such as the surge, sway, and roll motions.
- *Low-frequency (slow drift) cyclic loads* that excite the entire floating system including its mooring system at its natural periods in surge, sway, and yaw. Typical natural periods range from 3 to 10 minutes (180−600 seconds). In spar platforms, these forces can also induce dynamic excitation at the pitch and roll natural periods.

The motions of the floating structure in the six degrees of freedom are driven by the wave energy, and these in turn impact the mooring loads. The floater is usually designed so that the natural periods of critical motions avoid the period of maximum wave energy in order to minimize the floater motions and mooring line tensions. For example, since the peak wave energy period in the US Gulf of Mexico is around 15 seconds in the design condition, the heave natural periods of spars and semisubmersibles are all kept above 20 seconds to minimize the heave motion.

## 3.2   Wind load

Wind imposes static and low-frequency dynamic loads on the floating structure. It also generates waves and currents that add loads to the structure and mooring lines.

### 3.2.1   Description of winds

Wind is typically defined by direction, speed, and spectrum. Wind direction is usually defined as the direction from which the wind blows, thus a "northerly" wind blows from the north and goes to the south. The wind speed varies with time, as shown in Fig. 3.2. It also varies with the height above the sea surface. Fig. 3.3 shows an example of wind profile as a function of time average and elevation. For these reasons, the wind speed is defined by a time average at a reference elevation above the sea level. A commonly used reference height is 10 m. Commonly used averaged times include 1 minute, 10 minutes, and 1 hour. Wind speed averaged over 1 minute is often referred as sustained wind speed.

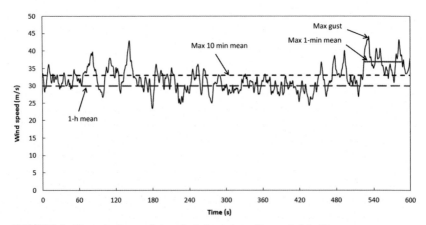

**FIGURE 3.2**   Example time variation of wind speed at reference height 10 m.

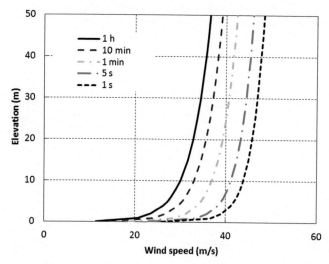

**FIGURE 3.3**  Example wind profile as a function of time average and elevation (Based on the NPD spectra, 30 m/s reference wind).

Fluctuating wind can be modeled by a steady component, based on the 1-hour average velocity, plus a time-varying component calculated from a suitable empirical wind gust spectrum. A number of wind spectra have been developed from various resources, such as the Ochi, Davenport, Harris, American Petroleum Institute, and norwegian petroleum directorate (NPD) spectra [2]. Currently the NPD spectrum [3] is most commonly used by the offshore industry.

The sustained wind velocity profile varies with height above the sea surface. It can be expressed as a power function:

$$U(z) = u_{10} \times \left(\frac{z}{10}\right)^b \tag{3.1}$$

where $U$ is wind speed; $u_{10}$ is wind speed at 10 m above sea surface; $z$ is height above sea surface; and $b$ is power of wind profile, typically 0.125.

There are other types of wind profile models applied in the offshore industry, including logarithmic profile model and others.

Of particular interest is the *squall*. A squall is a sudden wind storm that is usually associated with an active weather event, such as rain showers or thunderstorms. Squalls are typically of short duration (less than 2 hours), and are independent of wave and current conditions prior to the squall. Additionally, due to their short duration, squalls do not generate significant surface currents or waves at the site. A squall cannot be represented by the conventional approach using the wind spectrum. Therefore the wind speed and direction of a squall event are usually presented in the time domain as a time series. For mooring design work, a squall time series may be "scaled" such that the peak wind speed in the time history is equal to the squall wind speed of a specified return period (e.g., 10- or 100-year return period). The return period wind speeds are usually determined by site-specific extreme value analysis.

Offshore environments where squalls are a dominant environmental force introduce a challenge for mooring design and analysis. The squall winds are transient because they can rapidly change speed and direction. This makes it challenging to identify the maximum response for design, especially for a turret moored floating production storage and offloading (FPSO), which may change heading continuously under such an event. For more information on how to treat squall winds in a mooring design, refer to technical papers on this subject [4–6].

## 3.2.2  Wind-induced forces

The wind load characteristics of a floating structure are described by its wind force coefficients, which are obtained either by wind tunnel tests or numerical computations. They are in the format of a range of curves that give the wind load per unit wind velocity for all directions. With these curves, the wind loads including both the steady-state and dynamic components can be calculated.

The instantaneous wind force on a floating structure can be calculated by summing the instantaneous force on each member above the water line. This is calculated by an appropriate equation as shown in the following equation:

$$F_w = \frac{1}{2} \rho_a C_s A (V_Z + u' - \dot{x}) \left| V_Z + u' - \dot{x} \right| \tag{3.2}$$

where $F_w$ is the wind force; $\rho_a$ is the mass desity of air; $C_s$ is the shape coefficient (may also account for shielding); $A$ is the projected area of object; $V_Z$ is the mean wind speed; $u'$ is the instantaneous speed variation from sustained wind; and $\dot{x}$ is the instantaneous velocity of structural member.

For all angles of wind approaching the structure, forces on flat surfaces can be assumed to act normal to the surface, and forces on vertical cylindrical objects can be assumed to act in the wind direction. Forces on cylindrical objects that are not in a vertical attitude can be calculated using appropriate formulas that take into account the wind direction in relation to the attitude of the object. Forces on sides of buildings and other flat surfaces that are not perpendicular to the wind direction can also be calculated using appropriate formulas that account for the skewness between the wind direction and the plane of the surface.

In time domain calculations, time histories of wind velocities corresponding to wind spectra can be used in combination with the force calculations given in Eq. (3.2) to establish time histories of the wind forces. When using the frequency domain approach, it is common to linearize the force for spectral and frequency domain calculations, as shown in the following equation:

$$F_w = \frac{1}{2}\rho_a C_s A V_Z^2 + \rho_a C_s A V_Z u' \qquad (3.3)$$

where $F_w$ is the wind force; $\rho_a$ is the mass density of air; $C_s$ is the shape coefficient (may also account for shielding); $A$ is the projected area of object; $V_Z$ is the mean wind speed; and $u'$ is the instantaneous speed variation from sustained wind.

The first term of Eq. (3.3) is the steady wind force. $V_z$ corresponds to the mean wind speed used in generating the wind spectrum. Software programs are available for estimating the steady wind forces on a complex structure that contains many small dimensioned and spatially separated elements. The computed wind force is highly dependent on the modeling parameters entered in the program, and therefore it is essential for the users to have a good understanding of principles of wind tunnel tests when using these programs.

The fluctuating wind force may be calculated in the time or frequency domains. In the time domain, the total wind force is calculated from a time series of the instantaneous total wind velocity using Eq. (3.1). In frequency domain calculations, Eq. (3.3) is used with the wind spectrum to derive the wind force spectrum.

The total wind force on the floating structure may also be calculated using a simplified method. The total exposed area of the structure is multiplied with appropriate coefficients determined by model tests or computational numerical methods. In lieu of full spectral analysis, a quasistatic analysis using 1-minute mean wind speed and the first term of Eq. (3.3) can sometimes be used.

Wind and current forces for large ship-shaped structures can also be estimated based on data in the report published by Oil Company International Marine Forum (OCIMF) [7]. These simplified analytical tools were developed primarily for the analysis of mobile moorings. They may be used for preliminary designs of permanent moorings if more accurate information is not yet available at the early stage of the design process.

The shape coefficients in Table 3.1 can be used for perpendicular wind approach angles [8]. Shapes or combinations of shapes which do not readily fall into the specified categories will need special considerations.

Shielding coefficients may be used when the proximity of a second object relative to the first is such that it does not experience the full effect of the wind. More detailed information on wind force calculation can be found in [8–10].

**TABLE 3.1** Shape coefficients for calculating wind forces.

| Shape coefficient | Object |
| --- | --- |
| 0.4 | Spherical |
| 0.5 | Cylindrical shapes (all sizes) |
| 1.0 | Hull (surface type) |
| 1.0 | Deck house |
| 1.0 | Under deck areas (smooth surfaces) |
| 1.2 | Wires |
| 1.25 | Rig derrick (each face) |
| 1.3 | Under deck areas (exposed beams and girders) |
| 1.4 | Small parts |
| 1.5 | Isolated Structural shapes (cranes, angles, channels, beams, etc.) |

## 3.3    Wave load and vessel motions

Waves generate wave loads on the floater that include wave frequency dynamic loads and the slowly varying (low-frequency) wave drift forces. Most importantly, waves cause the floating structures to move in six degrees of freedom.

### 3.3.1    Description of waves and swells

There are two types of waves, that is, wind waves and swell. Wind waves, or wind-generated waves, are surface waves that result from the wind blowing over an area of water surface (i.e., fetch) [11]. Wind waves range in size from small ripples to large waves over 100 ft (30 m) high [12]. Wind waves are generated by the immediate local wind, and therefore they are also referred to as local waves or sea waves.

A swell is generated by global weather systems where wind blows for a duration of time over a fetch of water. A swell consists of waves of a long wave length that are not significantly affected by the local wind. They have been generated elsewhere from a distance some time ago. The directions of waves and swell are typically defined as the direction from which they are coming. A swell is featured with long wave periods (up to 20 seconds or more in some parts of the world), and is generally described as a symmetric narrow-band Gaussian spectrum.

The main parameters associated with waves are (Fig. 3.4):

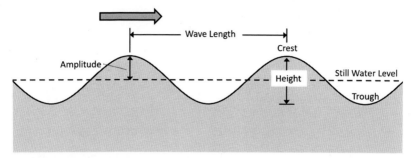

**FIGURE 3.4** Main dimensions associated with waves.

- wave height: vertical distance from trough to crest,
- wave length: distance from crest to crest of two consecutive waves in the direction of propagation,
- wave period: time interval between arrival of consecutive crests at a stationary point, and
- wave propagation direction.

Waves in a given area typically have a range of heights. For weather reporting and for scientific analysis of wind wave statistics, their characteristic height over a period of time is usually expressed as *significant wave height*, commonly denoted as $H_s$. This term represents an average height of the highest one-third of the waves in a given time period (usually chosen somewhere in the range from 20 minutes to 12 hours) or in a specific wave/storm system. The significant wave height can also be the value a trained observer, for example, a ship crew, would estimate from visual observation of a sea state. Given the variability of wave height, the maximum wave height, $H_{max}$, is approximately 1.6−2.0 times the significant wave height, $H_s$.

For regular waves, there are different wave theories. The simplest and most widely applied wave theory in offshore engineering is the linear wave theory, which is obtained by taking the wave height to be much smaller than both the wave length and the water depth. This theory is also referred to as small amplitude wave theory, sinusoidal wave theory or Airy theory. There are other wave theories for specific situations such as in very shallow waters, including second-order and higher order Stokes waves, cnoidal wave theory, and solitary wave theory [13,14].

Real ocean waves are irregular and random in shape, height, length, and speed of propagation. For mooring design, linear wave theory is usually used, and a linear random wave is modeled as the sum of many small linear wave components with different amplitudes and frequencies, Fig. 3.5 shows an example of superposition of regular waves. The wave phases are random with respect to each other to represent the random waves. With this simplification, the waves can be represented by an energy spectrum which gives the distribution of wave energy among different wave frequencies and heights on the sea surface.

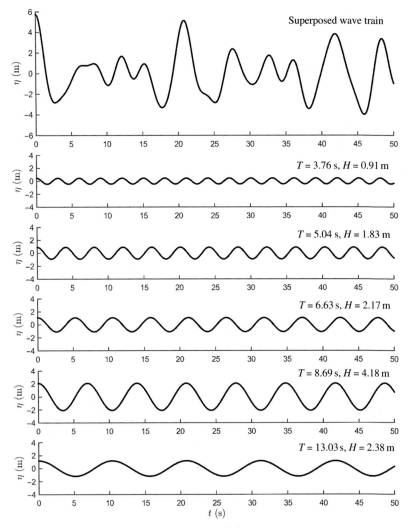

**FIGURE 3.5**   Superposition of regular waves to make irregular waves.

Wave spectrums are used to compute the mean wave drift force, the wave frequency motion responses, and the slow drift motions of the floater.

The linear random wave environment is often mathematically represented by various idealized wave spectra. There are many well established wave spectra available and the most commonly used ones are Joint North Sea Wave Observation Project (JONSWAP) spectrum and Pierson–Moskowitz spectrum. These wave spectrums are defined by the significant wave height, peak wave energy period, and the spectrum narrowness factor. Perhaps the simplest spectrum is the one proposed by Pierson and Moskowitz [15] as shown in Fig. 3.6. They assumed that if the wind blew steadily for a long

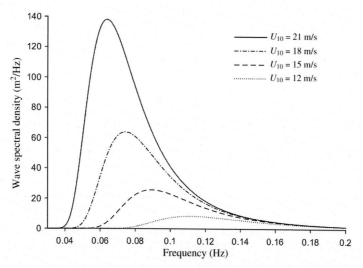

**FIGURE 3.6**  Pierson—Moskowitz spectra of a fully developed sea for different wind speeds.

time over a large area, the waves would come into equilibrium with the wind. This is the concept of a fully developed sea, that is, a sea produced by winds blowing steadily over hundreds of miles for several days.

After analyzing data collected during the JONSWAP, Hasselmann et al. [16] found that the wave spectrum is never fully developed. It continues to develop through nonlinear, wave—wave interactions even for very long times and distances. Hence an extra and somewhat artificial factor was added to the Pierson—Moskowitz spectrum in order to improve the fit to their measurements. The JONSWAP spectrum is thus a Pierson—Moskowitz spectrum multiplied by a function of peak enhancement factor (Fig. 3.7A). A wave height time history as shown in Fig. 3.7B can be transformed to a wave spectrum, which is then approximated by the JONSWAP spetrum.

### 3.3.2  Wave-induced forces and motions

Different wave force regimes are applied depending on a structure member's characteristic dimension. For slender structural members, such as mooring lines or risers, wave loads may be calculated using Morison's load formula. For large-volume hull structures, the wave radiation—diffraction approach is applied to compute the wave loads.

Model tests or numerical computations are used to obtain wave-induced motions and loads for large-volume floating hulls that are inertia-dominated with respect to global motion behavior. Full-scale field tests may be desirable, but they are simply too expensive and difficult to perform under

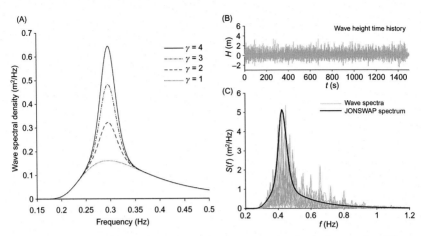

**FIGURE 3.7** JONSWAP Spectra. (A) JONSWAP spectra of a developing sea for different peak enhancement factor $\gamma$; (B) Wave height in time series; (C) Comparison between JONSWAP and the spectrum generated from the time series.

controlled conditions; and it may also be unrealistic to wait for the extreme weather to occur. Scaled model tests are therefore used. As discussed in Chapter 7, One drawback with model tests is the difficulty of scaling the test results to full-scale results when viscous hydrodynamic forces matter. The geometrical dimensions and equipment of the model test facilities may also limit the experimental possibilities.

Numerical computations have played an increasing role in calculating wave-induced motions and loads due to rapid development of computers. Determination of wave forces and motions can be made through the use of sophisticated hydrodynamic software programs, such as WAMIT, AQWA, and WADAM. Because of significant technology development in this area, these programs can provide good predictions on wave forces.

These numerical computations use radiation—diffraction theory based on the boundary integral equation method (BIEM) [17,18]. The basic idea of BIEM is to convert a three-dimensional (3D) problem in the whole fluid domain into a two-dimensional (2D) problem on the floater's surface. For this reason, the method is also referred to as the Boundary Element Method. The mathematical foundation of BIEM is Green's theorem that relates 3D volume integrals to 2D surface integrals. In contrast to the finite volume method that is used in computational fluid dynamics (CFD), BIEM has the advantage of reducing the dimensions by one and thus is faster in its computation. Since radiation—diffraction theory is based on potential theory without accounting for viscosity of the fluid, a Morison model of the floater is generally used for calculating the viscous drag force; this feature is built into most software programs.

Note, however, it may be unrealistic to expect that numerical computations will totally replace model tests in the foreseeable future. The ideal way is to combine model tests and numerical calculations. Results from model tests often give more confidence than those from numerical simulations, especially when novel concepts are tested out. Software programs do have the advantage that they can be run in a more efficient way than model tests to evaluate different floater designs in a large variety of sea conditions. However, interpreting and adjusting the computed results requires a physical understanding of the fundamental on wave-induced motions, which may be gained from experience with model tests and guidance in several references [13,14,18,19].

The purpose of wave diffraction analysis is to obtain the motion characteristics of the floater, listed below. They are the required input data for a mooring analysis:

- response amplitude operator, also known as motion transfer functions,
- mean wave drift force coefficients, and
- potential damping coefficients and frequency-dependent added mass.

To prepare for the wave diffraction analysis, the submerged hull of the floating structure is modeled by diffraction/radiation panels, and the mooring lines and risers are treated as Morison elements. The environmental parameters are input by the user in terms of characteristic period, wave spectrum, current velocity, wind spectrum, water depth, etc. To get the best result from the diffraction analysis, the wave period intervals near the natural frequencies of the floater motion should be refined to make sure that the peak of the resonant responses are well captured. In order to capture the characteristics of slow drift motions, a very long period close to the natural period of surge and sway motions should be added so that the added mass and potential damping to slow drift motions are accurately predicted.

Fig. 3.8 shows an example of a panel model and corresponding Morison model for a semisubmersible hull. The panel model includes four columns and a ring pontoon. Notice that only the wet surface below the sea level is modeled. This is because the integration over the fluid domain is only conducted on the still-water wet surface. As a result, it needs to be noted that geometry changes due to large heeling motions and wave run-up is not taken into account.

The Morison model, on the right-hand side of Fig. 3.8, has four single dots near the column at the pontoon level. These four points denote mooring elements. Mooring lines affect the motion of the floater, so they need to be accounted for accordingly. For a floater with mooring lines, the mooring lines will affect the stiffness, added mass, and damping of the floater, and they need to be modeled to capture the effects.

Another effect due to the existence of the mooring lines is the increased total floater displacement, which is equal to floater's mass plus vertical tension of the mooring lines. However, this presents a problem for the

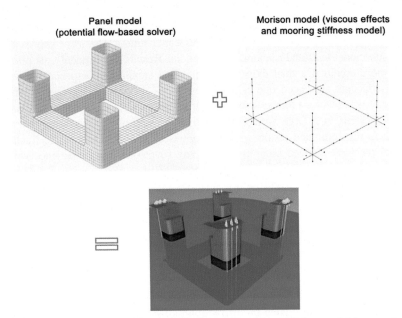

**FIGURE 3.8** Example of a diffraction/radiation panel model and Morison model for a semisubmersible hull.

motion analysis. If the actual mass of the floater is applied, the inertial force might be underestimated because parts of mooring lines also move with the floater. By using the total displacement to represent the inertial term, the calculation would overestimate the inertial term's impact on the floater's motion because not all parts of mooring lines move together with the floater. To resolve this problem, the concept of effective mass is adopted in common practice. The effective mass may be estimated by the floater mass plus a percentage of the total mooring hanging weight, for example, say 30%. The percentage value can be verified by either a coupled analysis or a model test.

## 3.4 Current load and vortex-induced motion

Currents generate loads on the floating structure and its mooring and riser systems. They also generate vortex-induced motions (VIMs) on floaters with a deep-draft cylindrical hull such as Spar or a deep-draft columned hull such as a semisubmersible.

### 3.4.1 Description of currents

Current is typically defined by direction and velocity profile with depth. Current direction is defined as the direction toward which the current flows, which is opposite to the definition of wind direction. Current profile defines the current speed at different depths below the surface.

Current is treated as a steady-state phenomenon. Currents generate drag and lift forces on submerged hull structures, moorings, and risers. Strong currents also interact with waves to alter the wave parameters and wave loads. The current velocity varies with water depth. Close to the water surface, the current velocity profile is stretched or compressed due to surface waves. For most applications the current velocity can be considered as a steady flow field where the velocity vector (magnitude and direction) is a function of depth.

Given the current speed profile, the current load characteristics of the floater are described by its current force coefficients, which can be derived either by computation or wind tunnel tests. They are in the format of a range of curves that gives the current load per unit current velocity for all directions. It is noted that the current action increases loads in general, but may also reduce loads in a mooring design. On one hand, it adds to the mean static load; but on the other hand, it acts as a drag to the slow drift motion to the floater which is beneficial to the moorings.

The operation of a floating structure is affected mainly by three types of currents: ocean (including loop current), tidal, and storm surge currents.

*Ocean currents* are the vertical or horizontal movement of both surface and deep water throughout the world's oceans. Surface currents are those found in the upper 400 m (1300 ft) of the ocean. Surface currents are mostly caused by wind due to friction as it moves over the water. This friction then forces the water to move in a spiral pattern, creating gyres. In the northern hemisphere, gyres move clockwise; while in the southern hemisphere, they spin counterclockwise. The speed of surface currents is the greatest closer to the ocean's surface and decreases at about 100 m (328 ft) below the surface. Deepwater currents, also called thermohaline circulation, are found below 400 m. Gravity plays a role in the creation of deepwater currents, but these are mainly caused by density differences in the water.

Of particular interest is the *Loop Current* in the Gulf of Mexico, which is a warm ocean current that flows northward between Cuba and the Yucatán Peninsula, moves north into the Gulf of Mexico, loops east and south before exiting to the east through the Florida Straits and joining the Gulf Stream. Serving as the dominant circulation feature in the Eastern Gulf of Mexico, the Loop Currents can reach maximum flow speeds of 1.8 m/s [20]. In the Gulf of Mexico, the deepest areas of warm water are associated with the Loop Current and the eddy currents that have spun off and separated from the Loop Current. It is interesting to note that the warm waters of the Loop Current and its associated eddies provide more energy to hurricanes and allow them to intensify.

Tidal currents are strongest in large water depths away from the coastline and in straits where the current is forced into a narrow area. The most important tidal currents in relation to coastal morphology are the currents generated in tidal inlets. Typical maximum current speeds in tidal inlets are approximately 1 m/s, whereas tidal current speeds in straits and estuaries can reach speeds as high as 3 m/s.

Storm surge current is the current generated by the total effect of the wind shear stress and the barometric pressure gradients over the entire area of water affected by a specific storm. This type of current is similar to the tidal currents. The current velocity follows a logarithmic distribution in the water profile and has the same characteristics as the tidal current. It is strongest at large water depths away from the coastline and in confined areas, such as straits and tidal inlets.

### 3.4.2 Current-induced forces and vortex-induced motion

Currents can cause two major effects on the floater, that is, drag force and vortex induced motion (VIM). Drag force can be on both the floater and its mooring lines. Motions caused by vortex shedding (i.e., VIM) are mainly a side-to-side motion of the floater that creates additional tensions in the mooring lines.

The drag force exerted on a bluff component of a floating structure by a current is proportional to the square of the current velocity. The drag force acts in the direction of the component of current that is normal to the component axis. If the instantaneous velocity of the structural member is negligible, the current drag force can be determined using the formula in following equations:

$$F_d = \frac{1}{2}\rho_w C_D A_C V|V| \tag{3.4}$$

where $F_d$ is the drag force; $\rho_w$ is the mass density of water; $C_D$ is drag coefficient; $A_C$ is the projected area exposed to current; and $V$ is the current velocity vector normal to the plane of projected area.

The current effect is not limited to producing steady drag forces. Current can also produce low-frequency excitation and damping. The damping effect is especially important in the low-frequency range where damping from other sources is small. The current-induced damping should be considered for all low-frequency motion analyses. The current drag coefficient of the hull should be obtained by model testing or numerical computations.

The total current force on the structure may also be calculated using the total exposed area of the structure with appropriate coefficients determined by model tests or some other appropriate methods. Current and wind forces for large tankers can be also estimated by using the data in the report published by OCIMF [7]. These simplified analytical tools may be used for preliminary designs of permanent moorings if more accurate information is not available at the early stage of the design process.

Current can cause VIM on the floater. It is a well-known phenomenon that cylindrical columns exposed to a current create alternating eddies, or vortices, at a regular period. Fig. 3.9 shows how these eddies appear in the downstream wake of four columns, causing the semisubmersible to oscillate

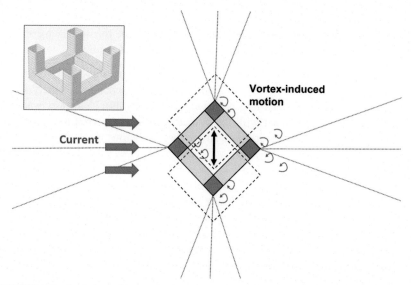

**FIGURE 3.9**   Semisubmersible experiencing VIM caused by a current. *VIM*, Vortex-induced motion.

in the direction perpendicular to the current. Similarly, floating structures consisting of a large diameter cylindrical body or columns such as spars and tension-leg platforms can experience low-frequency motions due to vortex shedding in the presence of currents [21]. These VIMs are most prominent for spars, where most of the industry experience has been acquired. Nevertheless, deep-draft semisubmersibles can also experience significant VIM, and this effect should be taken into account in the design [19].

VIM has three primary effects on the mooring design [2]:

1. The in-line drag coefficient is increased due to the presence of VIM.
2. The low-frequency VIM increases the mooring line tension.
3. Fatigue damage is incurred by the low-frequency tensions in mooring lines.

These effects should be taken into account in the strength and fatigue analyses of the mooring system. The occurrence of the loop current and associated eddies in the Gulf of Mexico, which can affect a particular site for an extended period of time, makes consideration of VIM particularly important for this geographic area. Note that prediction of VIM is typically done using model tests, but technology development has made the use of CFD very promising for VIM prediction in recent years.

## 3.5   Ice load

Ice generates loads by colliding with and pushing a floating structure. Ice loads are either completely avoided or somewhat mitigated by different ice management techniques depending on the type of ice features.

### 3.5.1 Description of ices

There are several types of ice features. Among them, the following three types may be more of a threat to any moored floating structures.

- *Iceberg*—An iceberg is a large piece of freshwater ice that has broken off a glacier or an ice shelf and is floating freely in open water. It may subsequently become frozen into pack ice. As it drifts into shallower waters, it may come into contact with the seabed, a process referred to as seabed gouging by ice. About 90% of an iceberg is below the surface of the seawater.
- *Level (and pack) ice*—Level ice is sea ice that has not been affected by deformation and is therefore relatively flat. It can be defined as sea ice with undeformed upper and lower surfaces. Level ice normally has a granular top layer, a transition zone, and then columnar grains through most of the remainder of the thickness of the ice sheet. Driven by winds and ocean currents, pack ice is a mixture of ice fragments of varying size and age that are squeezed together and cover the sea surface with little or no open water. At maximum expansion during the winter, pack ice covers about 5% of Arctic waters and about 8% of Antarctic waters.
- *Ridged ice*—Ridges are linear features created by the interaction between ice sheets/floes. Wind and currents can pile up level or pack ices to form ridges. The ice cover can deform when subjected to movement caused by external stresses on the ice sheet/floe and interactions with another ice sheet/floe.

Any of these ice features can be challenges for mooring operations, which have been addressed through a limited number of floating drilling and production operations in the past. Because the Arctic environment is frequently subject to large fluctuations in seasonal and year-to-year conditions, long-term observations are needed to understand the range of variability. The occurrence and geometry of specific ice features, such as icebergs, pack ice, and ridged/rafted ice should be determined on the basis of field measurements at the site or historical data from nearby sites. Statistics, such as frequency, probability distribution, and dimensions of the ice feature, should be determined for the purpose of selecting the type of the mooring system and designing it.

### 3.5.2 Ice-induced forces and ice management

For various ice features, ice loads on floating structures can be predicted by using the simplified procedures in ISO 19906 [22], which is the most recognized code for guidance on ice load predictions. It provides a formula that can be efficiently implemented in parametric studies. It can be used as a convenient starting point for ice load calculation. Then, the calculated ice load can be compared with the ice loads predicted by other methods such as full-scale measurement, model testing, or numerical simulation to arrive at an ice load for mooring design. Available full-scale measurement data are very limited, and the cost for collecting the data is very high. The use of model testing for ice load prediction is increasing; however properties of ice features and vessel are difficult to model due to scale effect. Numerical simulation is a promising technology, but it may need validation with model test or full-scale data.

When designing a mooring system for ice-prone areas, attention should be paid to the arrangement of mooring components [23]. For instance, mooring lines should be routed so as to avoid direct exposure to ice actions in the splash zone and below, depending on the design of ice interaction scenarios. Fairleads should be positioned to minimize such effects or localized ice management may be adopted. It should be understood that ice features caught by the mooring lines can result in damage to the mooring components.

The mooring system is usually designed to resist the load of small pack ices, but not the impact loads of icebergs and large ridged ices. Therefore floating systems are better off designed not to interact with ice features such as icebergs, large ridged ices, or large pack ices. This is simply because mooring lines have very limited load capacity. It is usual to employ ice management [24,25], such as deploying an ice-breaking vessel to break the surrounding level/pack ice or sending a tug boat to tow an iceberg away. There was a technique established based on the ice management experience gained from floating drilling operations in the Beaufort Sea [23,25]: a large ice-breaker upstream of the moored vessel breaks the ice into large fragments, and smaller icebreakers break those fragments into smaller ones that flow around the moored vessel [24].

Another solution is to adopt a disconnectable mooring system which allows the floating production vessel to sail away from any incoming ice feature, particularly icebergs, large ridged ices, or large pack ices. Refer to Chapter 2, Types of mooring systems, for details on disconnectable turret systems. In order to avoid any production shutdown due to a disconnection, an ice management plan is also developed in addition to adopting a disconnectable mooring system. The plan may involve a combination of detection, monitoring, ice-breaking, and ice-towing within the context of alert zones, which require certain procedures to safeguard the production/drilling wells. There have been a couple of disconnectable mooring systems for FPSOs installed in an ice-prone region (i.e., Terra Nova and White Rose) off the Canadian Grand Banks east of Newfoundland [26].

## 3.6   Other topics on environment loads

In this section, some specific aspects of environmental loads are briefly introduced, and if interested, the readers are encouraged to do more background reading.

### 3.6.1 Directional combination of wind, waves, and current

For semisubmersibles and spars with spread moorings, the collinear condition is usually the most critical combination of environment directionality which governs the mooring system design. For spread-moored FPSO systems, the collinear and near-collinear combinations with wind, waves, and current coming from the beam direction are usually the most critical and shall be carefully evaluated.

For FPSOs with single-point mooring systems, the collinear condition may not be the governing condition, and therefore both the collinear and noncollinear conditions should be considered in the mooring design. Ideally, the directional combination should be derived based on site-specific design data. However, in reality, such directional site-specific data are usually not available. In such cases, the combination of environmental directions should be developed including at least the following:

- collinear condition—collinear wind, waves, and current and
- cross condition—wind at 30 degrees and current at 45 degrees from waves.

For all the combinations, the environment should be rotated with reference to the mooring line direction by a 15-degree interval to ensure that all the possible scenarios are covered.

It is noted that the current practices of directionality combination of environments recommended by class societies such as American Bureau of Shipping, Det Norske Veritas, and Bureau Veritas (BV) are different. For example, BV recommends the combination of current with large offset angles with wind and waves, and introduced load reduction factors.

### 3.6.2 Sensitivity study on wave period

In a typical design basis, the design waves are defined by the significant wave height, the peak energy period, and the wave spectrum. However, the relationship between the significant wave height and peak period is by no means fixed. What is defined in the design basis is most likely the most probable wave period associated with the significant wave height. For certain types of mooring systems, the wave period may have more impact on dynamic responses than the wave height.

In order to capture the most critical design condition for the mooring system design, wave period variation should be seriously considered. The most rigorous approach is to analyze the mooring system dynamics for all possible combination of wave height and period, which often means moving along the 100-year return wave height and period contour. This approach is very time-consuming and is only recommended for the detailed design stage. Alternatively, in a simplified method, the mooring system should be analyzed at least with $\pm 10\%$ variation of peak wave energy period.

### 3.6.3    Wave–current interaction

It has long been known that wave drift forces are affected by the speed of current. In the presence of strong current, the quadratic transfer functions (QTFs), which gives the second-order response in irregular sea, should be corrected to reflect the influence of current. In some cases, such an influence can significantly affect the mean wave drift forces and the slow drift motion of a floater.

Aranha [27] has expressed the wave drift damping as a function of the diagonal term of the QTFs at zero speed and of the forward speed of the floater. Similarly, a current, which can be seen as a relative forward speed of the vessel, induces a modification of the wave drift force. It has shown that the wave–current interaction has an effect on the wave drift force for floaters.

## 3.7    Questions

1. What are the three most important environmental loads to be considered when designing a mooring system?
2. Explain what is a wind spectrum? Name two wind spectrums that are commonly used by the offshore industry.
3. Besides model testing, what is the other commonly used method for evaluating wave-induced motions for a floating structure? After evaluating the vessel motions, what data are obtained that can be used as inputs for a mooring analysis?
4. What kind of environment causes VIMs for a floating structure? What types of structure are most susceptible to VIM?
5. Name at least two types of ice features. Briefly explain how they can impact a mooring design.

## References

[1]  API, Derivation of Metocean Design and Operating Conditions, first ed., API RP 2MET, 2014.

[2]  API, API RP 2SK Recommended Practice for Design and Analysis of Stationkeeping Systems for Floating Structures, third ed., American Petroleum Institute (API), 2005.

[3]  Norwegian Technology Standards Institution, NORSOK Standard: Actions and Effects: N-003, Rev.1, Norwegian Technology Standards Institution, Oslo, 1999.

[4]  Z. Zhong, L. Yong, C. Dusan, F(P)SO global responses in the west of Africa squall environment, in: ASME 2005 24th International Conference on Offshore Mechanics and Arctic Engineering, American Society of Mechanical Engineers, 2005.

[5]  A. Duggal, et al, Response of FPSO systems to squalls, in: ASME 2011 30th International Conference on Ocean, Offshore and Arctic Engineering, American Society of Mechanical Engineers, 2011.

[6]  F. Legerstee, et al, Squall: nightmare for designers of deepwater West African mooring systems, in: 25th International Conference on Offshore Mechanics and Arctic Engineering, American Society of Mechanical Engineers, 2006.

[7] OCIMF, Prediction of Wind and Current Loads on VLCCs, second ed., Oil Companies International Marine Forum (OCIMF), 1994.

[8] ABS, Rules for Building and Classing Mobile Offshore Drilling Units. Part 3 Hull Construction and Equipment, 2016.

[9] API, Planning, Designing, and Constructing Tension Leg Platforms, third ed., API RP 2T, 2010.

[10] G.L. Dnv, Recommended Practice Environmental Conditions and Environmental Loads, DNVGL-RP-C205, 2017.

[11] I.R. Young, Wind Generated Ocean Waves., Elsevier, 1999. ISBN 0-08-043317-0.

[12] H.L. Tolman, in: M.F. Mahmood (Ed.), CBMS Conference Proceedings on Water Waves: Theory and Experiment (PDF), Howard University, Washington, DC, 13−18 May 2008, World Scientific Publications, 2010. ISBN 978-981-4304-23-8.

[13] C. Mei, ISBN 9971-5-0773-0 The Applied Dynamics of Ocean Surface Waves., World Scientific, Singapore, 1989.

[14] C. Mei, ISBN 0-521-58798-0 Mathematical Analysis in Engineering., Cambridge University Press, 1997.

[15] L. Moskowitz, Estimates of the power spectrums for fully developed seas for wind speeds of 20 to 40 knots, J. Geophys. Res. 69 (24) (1964) 5161−5179.

[16] K. Hasselmann, T.P. Barnett, E. Bouws, H. Carlson, D.E. Cartwright, K. Enke, et al., Measurements of wind-wave growth and swell decay during the Joint North Sea Wave Project (JONSWAP), Ergänzungsheft 12 (1973) 1−95.

[17] O.M. Faltinsen, Sea Loads on Ships and Offshore Structures, Cambridge University Press, 1990.

[18] J.N. Newman, Marine Hydrodynamics, MIT Press, Cambridge, MA, 1977.

[19] W. Ma, et al, Vortex induced motions of a column stabilized floater, in: Proceedings of the DOT International Conference, 2013.

[20] A. Gordon, Circulation of the Caribbean Sea, J. Geophys. Res. 72 (24) (1967) 6207−6223.

[21] D. Roddier, T. Finnigan, S. Liapis, Influence of the Reynolds number on spar vortex induced motions (VIM): multiple scale model test comparisons, in: ASME 2009 28th International Conference on Ocean, Offshore and Arctic Engineering, American Society of Mechanical Engineers, 2009.

[22] ISO, "Petroleum and Natural Gas Industries−Arctic Offshore Structures", ISO 19906, First Edition, 2010-12-15.

[23] C. Makrygiannis, R. McKenna, B. Wright, T. Sildnes, W. Jolles, M. Mørland, et al., Ice management and operational strategy for floaters in ice, in: Proceedings of the International Conference on Port and Ocean Engineering Under Arctic Conditions, no. POAC11-057, 2011.

[24] A. Palmer, Arctic Offshore Engineering., World Scientific, 2013.

[25] J. Hamilton, et al, Ice management for support of arctic floating operations, in: Proceedings of the Arctic Technology Conference, 2011.

[26] G.B. Howell, A.S. Duggal, G.V. Lever, The terra nova FPSO turret mooring system, in: Proceedings of the Annual Offshore Technology Conference, 2001.

[27] J. Aranha, Second-order horizontal steady forces and moment on a floating body with small forward speed, J. Fluid Mech. 313 (1996) 39−54.

# Chapter 4

# Mooring design

## Chapter Outline

4.1 **Design basis** 64
    4.1.1 Gather input data 64
4.2 **Design process** 65
    4.2.1 Select mooring system type 67
    4.2.2 Determine the profile (catenary or taut leg) 67
    4.2.3 Design the mooring pattern 68
    4.2.4 Design the mooring line composition 71
    4.2.5 Optimize the mooring design 73
4.3 **Design considerations** 74
    4.3.1 Limiting vessel offset 75
    4.3.2 Minimizing line tension 75
    4.3.3 Reducing fatigue damage accumulation 76

    4.3.4 Avoiding clash or interference 76
4.4 **Design criteria** 77
    4.4.1 Design codes 77
    4.4.2 Vessel offset requirement 77
    4.4.3 Strength design criteria 78
    4.4.4 Fatigue design criteria 79
    4.4.5 Operability requirement 80
4.5 **Engineering analysis and code check** 80
    4.5.1 Mooring analysis load cases 81
4.6 **Questions** 82
**References** 83

Mooring design methodology is introduced in this chapter with the focus on moorings for mobile offshore units and permanent production facilities. Mooring system design has to comply with the regulations of the coastal country, industry standards, and class society rules (if applicable). In addition, the mooring system design also needs to meet the owner's (or project's) specific requirements.

The mooring design is an iterative process that requires an integrated systematic approach. The design process involves the following main steps:

1. Assemble the project specifications and floating system information.
2. Define environmental conditions and develop load cases.
3. Perform hydrodynamic analysis for the floater.
4. Perform mooring analysis on the preliminary mooring design.
5. Check codes and design criteria for compliance.
6. Select mooring components and equipment.

Mooring System Engineering for Offshore Structures. DOI: https://doi.org/10.1016/B978-0-12-818551-3.00004-1

This chapter introduces the design process, design considerations, and typical design criteria. Details of mooring analysis are introduced in the next chapters.

## 4.1   Design basis

When a new design project is started, the owner/operator (i.e., client) prepares a design basis for each engineering discipline, such as hull, mooring, and riser, etc. The design basis, also known as design premise, contains all the essential information and requirements for a chosen engineering contractor to follow. The contractor's mooring engineers start the process of mooring design based on the design basis. An outline of a sample design basis is given below:

- Mooring system design information
  - Location and water depth
  - Vessel geometry and loading conditions
  - Metocean conditions and design load cases
  - Risers, umbilicals, and flowlines information
  - Design analysis software to be used
- Mooring system design constraints
  - Layout of subsea infrastructure such as pipelines
  - Map of seabed hazards such as sensitive marine habitats
- Mooring system design criteria
  - Design life
  - Design standards
  - Vessel offset limit
  - Strength criteria
  - Fatigue criteria
  - Corrosion allowance
  - Anchor design requirement
- Deliverables

While the design basis contains many items, the main objectives of mooring design can be summarized as the following:

1. To maintain the floating structure on station within specified tolerances under normal operating and extreme storm conditions; and the excursion of the floater must be kept within the limit without overstretching the risers and umbilicals;
2. To provide the mooring system with sufficient strength and fatigue life to guarantee the operability and reliability of the offshore system.

### 4.1.1   Gather input data

To start the mooring system design, the basic input data need to be gathered which include the characteristic parameters of the floating structure, the design environmental conditions, and the floater's operation requirements. These input parameters can be broadly categorized into the following groups:

1. Nature of operation
   - Type of operation; manned or unmanned operation
   - Duration or design life
2. Environmental conditions
   - Water depth and seabed profile
   - The extreme (maximum design) and operating conditions for wind, wave/swell, and current and their directionality. Table 4.1 gives an example
   - Long-term distribution of environmental conditions including wave scatter diagram
   - Marine growth
3. Floater properties
   - The key dimensions of the floating structure including the lines plan that defines the body profile
   - The floater displacement, position of the center of gravity, weight, moment of inertia for all relevant drafts
   - The layout of the upper deck structure, and the superstructure. The sizes and specific locations of deck equipment and buildings that define the wind areas
   - For a turret system, the location and size of the turret
   - Risers' properties
4. Site information
   - Geotechnical information, for example, soil properties derived from core samples
   - Geographical location
   - Existing installations or infrastructure on the surface and subsea

## 4.2 Design process

For a floating structure to be deployed in a specific offshore field location, its mooring system design is tailor made, taking into account the floater characteristics, field characteristics, extreme environments, mode of operations, etc. The design process for a new mooring system involves the definition of the following aspects:

- Type of the mooring system
- Mooring profile
- Mooring pattern (including anchor radius, number of lines, and spread angles)
- Mooring line composition
- Type of anchor
- Onboard equipment

**TABLE 4.1 Example of design environments.**

| | | | Units | 100-year wave driven | 100-year wind driven | 100-year current driven | 1-year return |
|---|---|---|---|---|---|---|---|
| Wave | Wave spectrum | | | JONSWAP | JONSWAP | JONSWAP | JONSWAP |
| | Gamma | | | 3.3 | 3.3 | 3.3 | 3.3 |
| | Sig. wave height | | m | 12.9 | 12.5 | 11.7 | 7.7 |
| | Maximum wave height | | m | 21.5 | 20.8 | 19.5 | 12.9 |
| | Peak wave period | | s | 16.1 ± 2 | 15.6 ± 2 | 14.6 ± 2 | 13.4 ± 1.5 |
| Wind | Wind spectrum | | | | | | |
| | 1-h mean | | m/s | 39.4 | 42.6 | 38.2 | 30.40 |
| | 1-min mean | | m/s | 49.6 | 53.6 | 48.2 | 38.30 |
| Current | Surface speed | | cm/s | 161 | 172 | 209 | 105 |
| Depth factor | 0.1 h | 100 | cm/s | 161 | 172 | 209 | 105 |
| | 0.2 h | 200 | cm/s | 138 | 148 | 179 | 93 |
| | 0.3 h | 300 | cm/s | 115 | 123 | 149 | 80 |
| | 0.4 h | 400 | cm/s | 98 | 104 | 126 | 70 |
| | 0.5 h | 500 | cm/s | 81 | 86 | 104 | 61 |
| | 0.6 h | 600 | cm/s | 70 | 75 | 91 | 56 |
| | 0.7 h | 700 | cm/s | 60 | 64 | 77 | 52 |
| | 0.8 h | 800 | cm/s | 53 | 56 | 67 | 49 |
| | 0.9 h | 900 | cm/s | 45 | 48 | 57 | 46 |
| | 1.0 h | 1000 | cm/s | 45 | 48 | 57 | 46 |
| Tidal | Maximum water level | | cm | 146 | 146 | 146 | 129 |
| | Minimum water level | | cm | −120 | −120 | −120 | −115 |

(Depth factor rows: the value column under "m" gives the depth in metres: 100, 200, 300, 400, 500, 600, 700, 800, 900, 1000.)

### 4.2.1    Select mooring system type

Selecting the mooring system type is the first step in the mooring system design. As described in Chapter 2, Types of mooring systems, mooring systems can be classified into spread and single-point moorings. The spread mooring system that fixes the floater heading is most widely used for mooring of nonship-shaped floating structures such as semisubmersibles and spars. The single-point mooring system is usually adopted for ship-shaped structures with slender body configuration. Note that the spread mooring system can still be designed for ship-shaped structures in relatively calm environments with highly directional wind, waves, and current; otherwise, the turret single-point mooring system is normally selected.

As discussed in Chapter 2, Types of mooring systems, disconnectability is another design choice. In most cases, the mooring system is permanently connected with the floater, that is, disconnection is not a planned operation. In special situations, the mooring system is designed to be disconnectable, especially those for floating production storage and offloadings in harsh environments with icebergs or tropical cyclones.

Most mooring systems are designed to be passive. That is, no human intervention is required during normal operations. However, a mooring system can be designed to accommodate active human intervention by either using thrusters or onboard winches.

### 4.2.2    Determine the profile (catenary or taut leg)

The most common types of mooring profiles used for floating production systems are as follows (Fig. 4.1):

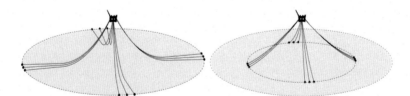

**FIGURE 4.1**    Catenary and taut leg mooring systems.

1. Catenary system with an *all-chain* setup (best for very shallow waters).
2. Catenary system with a *chain−wire−chain* setup.
3. Taut (or semitaut) leg system with a *chain−wire−chain* setup.
4. Taut (or semitaut) leg system with a *chain−polyester−chain* setup (best for ultradeep waters).

The selection of a technically feasible and cost-effective mooring system type is primarily dependent on water depth and environment. Additional factors to be considered are vessel offset restrictions imposed by risers and umbilicals. Further guidance for selecting a mooring profile is provided below:

- *For water depth less than 500 m*—Catenary systems are the most cost-effective choice. Both all-chain and chain—wire—chain should be considered. The latter may be more cost-effective for depths larger than 300 m.
- *For water depth between 500 and 1000 m*—All four choices may be considered. Offset constraints imposed by riser type or vicinity to other structures may govern selection.
- *For water depth between 1000 and 2000 m*—Taut leg systems are the most cost-effective choices. Both chain—polyester—chain and chain—wire—chain should be considered.
- *For water depth greater than 2000 m*—Taut leg system with polyester rope is likely the most cost-effective choice, especially for harsh environments. A chain—wire—chain system can still be considered, but it gets heavier as the water depth increases.

Selection of mooring system profiles may involve a comparison of two or more kinds. In general, catenary systems require a larger *R/D* ratio (anchor radius/water depth) than taut leg systems. Catenary systems tend to be costlier at larger water depths (say, more than 1000 m), as the mooring lines become very long. Taut leg systems can be costlier in shallower waters, because the high stiffness of the short lines leads to very high tensions and, in turn, requires very large mooring component sizes.

The taut leg systems can provide considerably smaller vessel offset than the catenary systems. As such, they are typically required to allow the use of steel catenary risers. These risers demand a small vessel offset so that they do not experience local yielding. Flexible risers, on the other hand, can be designed to tolerate larger offset and can be combined with any of the mooring systems. Vessel offset in extreme conditions for taut leg systems are typically small at around 5% of water depth, while catenary systems have much larger vessel offsets.

## 4.2.3 Design the mooring pattern

The selection of a mooring system configuration is typically done by varying the parameters until a cost-effective system is found that complies with regulatory and functional requirements. Following are the key steps in the design process. The design steps below can be performed in any order. To reach a final configuration, it may take several iterations by varying the parameters in different combinations. Converging on an optimal mooring system design will usually consist of a "familiarization process" involving a number of cycles. This familiarization process is typically done during the early concept phase of a project (also known as pre-FEED phase).

*Step 1: Determine anchor radius (i.e., anchor distance)*—For a deepwater taut leg system, a good starting point for line length is to have a *R/D* ratio at 1.4, that is, anchor radius is 1.4 times the water depth. For shallower water catenary systems, the anchor radius (i.e., anchor distance or anchor scope) is generally much larger. The large radius is required in order to have sufficient chain length resting on the sea floor to achieve a catenary (weight) effect and ensure no uplift on the anchor. Fig. 4.2 provides a reference on the *R/D* ratio for different water depths based on data from recent projects.

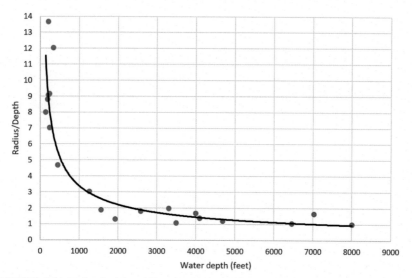

**FIGURE 4.2**  Ratio of anchor radius to water depth (i.e., *R/D* ratio) based on a number of existing facilities.

Note that the selection of the anchor locations and the layout of the mooring system are affected by several other factors, such as the geophysical profile of the field, soil composition characteristics at anchor locations, the layout of the subsea facilities, and directions of the metocean environments.

*Step 2: Determine number (and size) of lines*—This step consists simply of performing a qualified estimate of how many lines will provide a cost-effective mooring system. Typically, reference projects are used for the first design cycle. As a general rule, the number of lines should be kept to a minimum as this is likely to be the most cost-effective design due to fewer fairleads/winches and less installation time. The number of lines is increased if the required line size for a given number of lines becomes exceedingly large. Chain size available on the market is, at present, limited to approximately 220-mm diameter.

This step also determines the minimum line size for the given number of mooring lines. Mooring line size has to be checked iteratively by running mooring analyses. Pretensions of the lines are also checked to ensure they stay roughly between 10% and 20% of minimum breaking load (MBL). Note that pretension is the tension in the mooring lines in a zero environment (no wind, no wave, and no current). Mooring engineers like to keep pretensions as low as possible while still meeting the vessel offset requirement. Deepwater systems with steel wire ropes will likely see a higher pretension than systems with polyester ropes due to the higher submerged weight of steel wires. Systems requiring small vessel offsets will also get relatively higher pretensions.

Fig. 4.3 shows alternative mooring patterns with different numbers of lines and sizes for a semisubmersible. Obviously, the $4 \times 1$ pattern will not be able to pass the code check if the safety factor for one-line damaged condition is required to be met. The $4 \times 1$ pattern may still be a viable option for some special installations with a low failure consequence, such as an unmanned weather station or a renewable energy platform. Mooring patterns with fewer lines, such as $4 \times 2$, are normally governed by the safety factor for the damaged condition. With more lines, the mooring designs are then governed by the safety factor for the intact condition, and the redundancy check may become trivial. Such a trend is rather intuitive, as losing one line has a lesser impact on a mooring system with a large number of lines.

**FIGURE 4.3** Various mooring patterns for a semisubmersible.

Note that if an 8-point mooring is to be used, mooring specialists normally choose an $8 \times 1$ evenly spread pattern over a $4 \times 2$ clustered pattern. This is because the unbroken adjacent line in a $4 \times 2$ will be left in isolation with little help from the rest of the six lines. When one line is broken in an $8 \times 1$ pattern, the adjacent lines may still have a chance to share the load. However, when one line is broken in a $4 \times 2$ pattern, the remaining line in the cluster will most likely break as well, because it will not have the strength to withstand the load alone [1].

*Step 3: Determine grouping and spread angles*—This step starts by identifying if there are constraints on the mooring pattern, for example, large riser corridors or nearby floating/subsea infrastructures. A floating production system can have either a spread or a grouped (clustered) mooring pattern, as shown in Fig. 4.4. The spread mooring system in this context means that the angles between all lines are similar, so they are also called "evenly spread" moorings. Grouped mooring systems have three or four groups with tightly spaced lines in clusters. They are also called clustered moorings.

For semisubmersibles, the grouping will have to consist of four groups, while spars can have three or four groups. The selection of grouping or not will be project specific. Generally, for a system with a large number of mooring lines (say, more than nine), grouped systems are preferred to allow more open space for the risers (as shown in Fig. 4.4) and better sharing of loads. For grouped systems, the spread angle between adjacent lines is typically 3−5 degrees.

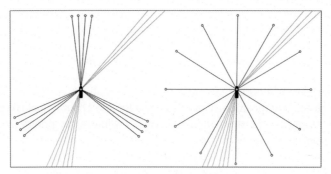

**FIGURE 4.4**    Grouped versus evenly spread mooring patterns.

Typical mooring designs often use a symmetrical spread of mooring lines, that is, all mooring lines are spread equally with a common angle between the mooring legs. The nontypical mooring design can group the mooring lines in bundles to create space for risers and umbilicals. In some cases, the designer would take advantage of the directionality of the environment or the riser bundle to lay out the mooring pattern in a nonsymmetric way. The use of a noneven spread or a nonuniform anchor radius can reduce the mooring system cost but may slightly increase the engineering and installation efforts.

Note the mooring line sector may be restricted by factors such as the subsea field layout, proximity to any nearby installations, sea-floor conditions, riser configuration and in some rare cases block lease boundaries. In selection of the mooring spread, the designer should also give consideration to the subsea facilities to avoid potential clashes with them.

### 4.2.4   Design the mooring line composition

A mooring leg includes several off-vessel components and on-vessel hardware that typically include the following from top to bottom:

- Mooring winches, chain jacks, chain stoppers.
- Fairleads.
- Chain, steel wire ropes or polyester ropes.
- Connectors (H-links, tri-plates, delta plates, shackles, ball-grab, insert, etc.).

- Buoys or clump weights (if applicable).
- Anchors.

The water depth is an important factor in determining the choice of mooring leg configuration and the mooring line material. The commonly used mooring line materials are chain, wire rope, and polyester as reviewed in Chapter 9, Hardware—off-vessel components, of this book. From the design point of view, choosing the line materials depends on the characteristics of the component, particularly the specific weight in seawater.

In general, *chain* is very heavy and costly but highly resistant to handling and abrasion. In shallow water, the heavy weight of the chain creates the so-called catenary (weight) effect that restricts the vessel offset and motions. Therefore, in shallow waters, the mooring line composition is often "all chain."

For permanent mooring systems, chain is typically used for the top segment in the line composition. This is because the top chain is locked securely by chain stoppers and stays there for years, whereas temporary moorings need to frequently run wire ropes with a good speed using linear winches (see Chapter 10: On-vessel equipment, for details). Special considerations have to be given to the *top chain length*, including the following:

- Length between the fairlead and the first connector. Minimum allowance at 10–15 m for chain repositioning (i.e., shifting of chain link periodically to rotate the wear point).
- Extra length to accommodate polyester rope elongation.
- Extra length to account for the tolerance of anchor location.
- Extra length to back-fill the segment length of a polyester test insert, if applicable.
- Extra length to allow the vessel to move an offset for drilling operations, if applicable.

*Wire rope* is less heavy than chain, and reasonably resistant to handling, so it is often used for the suspended portion of the mooring profile. In a shallow-water mooring system, it may also be laid down on the seabed to create a spring effect to ease the mooring line dynamic tension. Note, however, wire rope is not recommended to be placed at the touchdown point because it can suffer from excessive bending.

*Polyester rope* is typically used in deepwater moorings to reduce the mooring's self-weight and to absorb the line stretch under vessel motion. It is vulnerable to abrasion by the seafloor, so it is only used for the suspended portion in the mooring line composition. Polyester and wire segments should, in general, not touch the seabed in any normal or extreme conditions. Polyester ropes also need to stay submerged and maintain a clearance from the sea surface, say 100 m, to avoid any buildup of hard marine growth. Polyester test inserts are required for permanent mooring systems installed in the US waters [2]. Because polyester is considered by some as a relatively new technology, the test inserts can be retrieved for inspection and thus mitigate the uncertainty associated with the novelty. The length of test inserts is typically around 15−20 m.

Note that polyester ropes have the drawback of permanent elongation that comes from bedding-in and creep. Bedding-in of new ropes leads to an elongation of around 4%−6% of the rope length. The common practice is to remove as much as possible of the bedding-in elongation during installation by preloading (stretching) the lines. Creep is the increase in rope length under sustained tensions. Creep elongation generally is less than 1% of the rope length over the service life for a permanent mooring [3,4]. Winching-in of the top chain may be needed at a certain point in the service life to bring a mooring line back to its design length and tension.

The mooring leg is often made of segments of different materials. The mooring design involves the selection of the most suitable materials for different segments. The mooring line composition is determined, when the segments are defined in terms of segment length, size (diameter), and material.

In shallow-water moorings, clump weights and subsea buoys may be deployed. The application of clump weight is to increase the mooring system restoring force. When a floater is pushed by the environmental force, it will have to lift up the clump weights in the touch down segment before it gets moved (see Fig. 4.5). The application of mid-water buoys is typically for clearing seabed objects. Mid-water buoys can also be used to reduce payload on a submerged turret production buoy and to avoid touch down of wire ropes.

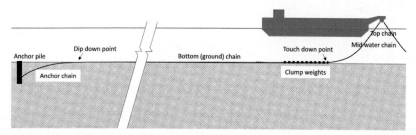

**FIGURE 4.5**   Mooring line profile showing clump weights in the touch down segment for a shallow-water mooring design.

## 4.2.5   Optimize the mooring design

Optimization of the mooring system is typically performed with respect to number of lines and size of the lines. Because a large percentage of the total mooring cost is incurred by the installation, any reduction in number of mooring lines can reduce the installation time and thus the cost. With fewer lines, the size of each mooring line will increase slightly, but the cost impact from the larger component size is normally small.

While the number of lines is one of the several parameters that can be optimized, there are other parameters that may be optimized as well. These include anchor radius, length of chain segment, pretension, etc.

The mooring system design process is highly integrated with mooring analysis. Safety factors, vessel offset, and anchor uplift angle have to be calculated by performing a mooring analysis for each design change. The design process is thus highly iterative. A large number of design and analysis cycles may be needed to reach a final design that is highly optimized.

Once the mooring design is optimized and finalized, the force—displacement (i.e., force—excursion) curve of the mooring system can be plotted, as shown in Fig. 4.6. The force—displacement curve is one of the important characteristics of the mooring system. It defines the relation between the vessel offset and the horizontal restoring force of the mooring system. Note that for polyester mooring systems, the polyester tensile stiffness changes over time, so the force—displacement curve is not a single line. Experienced mooring engineers often review the force—displacement curve to confirm that the mooring system is properly designed with a good station-keeping performance.

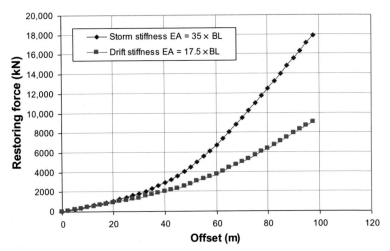

**FIGURE 4.6** Example of force—displacement curve of a polyester mooring system generated from its analysis model.

## 4.3 Design considerations

There are at least four variables to be tuned during the design process in order to develop an optimum mooring configuration that conforms to industry standards and class rules. These include vessel offset, line tension, fatigue damage, and clash avoidance.

## 4.3.1    Limiting vessel offset

Risers and umbilicals impose a limitation on the allowable vessel offset. In shallow waters, the ratio of floater extreme offset to water depth is much larger, and the risers/umbilicals are more likely to be overstretched. Therefore the floater must be kept in a smaller excursion radius. In deep waters, the risers/umbilicals feel less impact from the same offset and wave-induced vessel motions. This phenomenon partly explains why flexible risers are commonly used in shallow waters.

While the wave-frequency motion is independent to the mooring system, the vessel static offset and slow drift (i.e., low frequency) motion depend on many factors such as line profile, pretension, line material, mooring line spread, etc. In general, a taut leg profile tends to control the vessel offset better. Line material of high tensile stiffness can also reduce vessel offset. The possible means to reduce vessel offset include:

- Choose taut leg profile in deep water.
- Increase line pretension.
- Use more mooring lines.
- Use lightweight materials to minimize the catenary effect.
- Use clump weight or heavy chain at the touch down zone.
- Arrange line spread in the direction of extreme environment.

## 4.3.2    Minimizing line tension

The mooring line tension can be predicted by mooring analysis and if necessary verified by model tests. During the design process, mooring engineers increase the line diameter or adjust the mooring configuration until the tension safety factor meets the industry standards and class rules. In other words, the mooring line tension is kept within the design limit, which is 60% and 80% of MBL for the intact and damaged conditions, respectively, as required by most codes. Mooring engineers try to minimize line tension, because the tension in a mooring line can potentially cause various kinds of integrity issues, such as interlink wear and metal fatigue. However, reducing mooring tension often implies an increase in the material cost, so it is a tradeoff which is part of the mooring optimization. The possible means to minimize the line tension include:

- Select the most suitable mooring profile for the water depth.
- Use more mooring lines.
- Choose optimum line spread according to environment directionality.

- Choose lighter and less-stiff line material such as polyester.
- Use compliant mooring leg configuration.

### 4.3.3 Reducing fatigue damage accumulation

The field life is an important design parameter. The mooring system must have adequate fatigue life that exceeds the field life including fatigue safety factor, because fatigue has been one of the prominent failure modes (refer to Chapter 13: Mooring reliability, for details). Among the chain, wire, and polyester ropes, the chain is most vulnerable to fatigue failures. The possible means to improve the fatigue life include:

- Reduce dynamic tension—The fatigue damage increases with dynamic tension variation by a power of 3−5. Reduction of dynamic tension is an effective means of reducing fatigue damage. This can be done by improving the hull design so that vessel motion and vortex-induced motions (VIM) are reduced.
- Increase the line size—This is the most effective way to improve the fatigue life almost exponentially. However, an increase in tension variation may be a by-product of large line size, which may contribute to fatigue damage.
- Adopt a better design for fairleads—Out-of-plane bending fatigue can be mitigated by fairleads with dual articulations (double axis), low-friction bearings, and/or longer hawse pipes.
- Use other mooring materials than chain—Wire rope and polyester have better tension−tension fatigue endurance than chain.

### 4.3.4 Avoiding clash or interference

The mooring system often has to be designed in a congested field with many subsea facilities. As such, the mooring layout must accommodate the subsea facilities and flowlines. The mooring line should not clash with the risers, subsea facilities, or vessel hull. The mooring lines and risers should be designed so that they have a clearance in all environmental conditions. Crossing between the mooring lines and the risers is not permitted, especially with the mooring lines on top of the risers. Clashing between the mooring lines and the vessel hull is not permitted either. This can happen when the floater rolls and pitches heavily. It can also happen when the vessel overruns the mooring lines causing leeward legs to clash with the vessel bow. In addition, it is required that the anchors maintain a clearance (e.g., 150 m) to other facilities. The possible means to avoid clashing and to eliminate interference with subsea equipment include:

- Modify the spread angles between the mooring legs to provide more room for risers.

- Reduce anchor radius by using shorter and tauter mooring profile to raise the mooring lines up above any subsea facilities or pipelines.
- Insert buoys in the mooring line to raise or change the line profile.
- To prevent an external turret mooring system from clashing against the bow, cutoff the bow protrusion of the vessel or extend the turret supporting structure (cantilever).
- To prevent an internal turret mooring system from clashing against the bottom hull, increase the fairlead declination angle (with horizontal) by using heavy chain material or clump weight.

## 4.4   Design criteria

There are a number of well recognized codes and standards covering mooring design. It is often necessary to reference more than one source. In addition, there are project specific design requirements that should also be referenced. Care should be taken to use coherent design standards, input data, analysis methods, and safety factors. In some cases, particularly in relation to less-common types of mooring systems where design standards are not well established, the designer should use professional judgment to ensure the safety of the system.

### 4.4.1   Design codes

Common design codes used for mooring system design and analysis are briefly summarized below. The commonly used design rules and guidelines are [5,7–10]

- American Petroleum Institute Recommended Practice (API RP) 2SK "Recommended Practice for Design and Analysis of Stationkeeping Systems for Floating Structures."
- ISO 19901-7 "Petroleum and Natural Gas Industries—Specific Requirements for Offshore Structures, Part 7: Stationkeeping Systems for Floating Offshore Structures and Mobile Offshore Units."
- Det Norske Veritas (DNV) OS E301 "Position Mooring."
- American Bureau of Shipping (ABS) "Guide for Building and Classing Floating Production Installations."
- Bureau Veritas 493-NR "Classification of Mooring Systems for Permanent and Mobile Offshore Units."

The mooring design criteria typically define the minimum requirements for strength, fatigue life, and floater offset in relation to the design of risers and umbilicals.

### 4.4.2   Vessel offset requirement

The offset limit is specific to the project and is usually determined by the riser design. The floater's excursion has to be within the allowable limit to avoid overstretching the risers. An integrated design of the moorings and risers is therefore required to ensure that the mooring system design is compatible with the risers. In other words, the risers would not be overstretched when the floater offsets under environmental loading. In addition, during operation, there may be other structures and facilities around the floating structure, and therefore the offset of the floater in certain directions must be limited. In general, the floater would experience larger offset in deeper water as the mooring system is more compliant. However, when measured as a percentage of water depth, the offset percentage is larger in shallow waters.

### 4.4.3 Strength design criteria

The environmental parameters specifying the design condition should be developed from the metocean data for permanent or mobile moorings. The design condition is defined as that combination of wind, waves, and current for which the mooring system is designed. In practice, this is often approximated by the use of multiple sets of design conditions, such as the 100-year waves with associated winds and currents, the 100-year wind with associated waves and currents, and the 100-year current with associated wave and wind. The most severe directional combination of wind, wave, and current should be specified for the mooring system being considered. For permanent moorings, a return period of 100 years or higher should be used. The return periods used to account for environmental design conditions should be several times the design service life of the station-keeping system. Therefore a facility with a longer design life that is more than the typical 20 years may want to consider using a return period longer than 100 years.

For mobile moorings, the design environments with return periods of 5−10 years are used. When the mobile moorings are within the proximity of other installations during hurricane season, return periods of at least 10−20 years should be used depending on the level of close proximity, to account for the possible consequences of contact with surface, mid-depth or sea floor infrastructures or installations [5,6].

The strength safety factors specified by API RP 2SK are most commonly used for design of mooring systems. Note that mooring is one of those rare engineering disciplines that utilize a redundancy check [1]. The practice is to check if a mooring design can meet the first safety factor for the intact condition. Once that is proved positive, the design is further checked against a lower safety factor for the one-line damaged condition. Redundancy check is an effective second defense against any substandard mooring design, particularly those with fewer lines [1]. The required minimum safety factors are as shown in Table 4.2.

**TABLE 4.2** Safety factors required by API Recommended Practice 2SK.

| | Factor of safety for strength | |
| --- | --- | --- |
| | Using quasistatic analysis | Using dynamic analysis |
| Intact condition | 2.00 | 1.67 |
| One-line damaged condition | 1.43 | 1.25 |

Notice that higher safety factors are required if the quasistatic mooring analysis is used. In quasistatic analysis, the load on the mooring lines are calculated by statically offsetting the floater by wave-induced motions in the horizontal direction, while the dynamic responses of the mooring lines associated with mass, damping, and fluid acceleration are not considered. This method was commonly used in the early days of mooring design for moorings in relatively shallow waters. In dynamic analysis, the effects due to added mass, damping, fluid acceleration, and relative velocity between the mooring system and the fluid are all considered. The modern mooring designs are almost all based on the dynamic analysis and the associated tension safety factors should be used.

### 4.4.4   Fatigue design criteria

The principle of fatigue design is that the fatigue lives of mooring components should exceed the field life by appropriate safety factors. Since the fatigue analysis inherently deals with more uncertainty associated with T-N or S-N curves in comparison to the strength analysis, large safety factors (e.g., 3.0−10.0) are typically used as fatigue design criteria than those (e.g., 1.25−2.0) for strength design.

For permanent moorings, design conditions for fatigue consist of a set of environmental states that represents the long-term statistics of the local environment, taking into account the magnitude and direction of wind, wave, and current. VIM fatigue on mooring components is also assessed and included. For mobile moorings, fatigue assessment is not required, as the mooring components are replaced based on inspection results before they reach their fatigue limits.

If the mooring system is designed according to *API criteria*, the following criteria can be applied. The general format of $T-N$ curves is used for calculating nominal tension fatigue lives of mooring components. The fatigue *safety factor is 3.0* when the API criteria are used [5]. Note that ABS uses similar criteria as API, but increases the safety factor from 3.0 to 10.0 for mooring components that are noninspectable and in critical areas [7].

$$NR^M = K$$

where

N is number of cycles;

R is ratio of tension range (double amplitude) to reference breaking strength (RBS). For chain, RBS is taken as MBS (minimum breaking strength) of ORQ chain link of the same size for R3, R4, and R4S chain. For wire rope, RBS is the same as MBS;

M is slope of $T-N$ curve;

K is intercept of $T-N$ curve.

If the mooring system is designed according to *DNV criteria*, then the safety factor is calculated according to the fatigue damage ratio of adjacent lines. The following equation is used to calculate the mooring leg component fatigue capacity when DNV criteria are applied [8]. The fatigue *safety factor is between 5.0 and 8.0* when the DNV criteria is used (refer to Chapter 6: Fatigue analysis, for details).

$$N_c(s) = a_D \cdot s^{-m}$$

where

$N_c(s)$ is number of stress range to fail under stress $s$;

$s$ is stress range (double amplitude) in MPa defined as tension divided by the nominal cross-section area;

$a_D$ is the intercept parameter;

$m$ is the slope parameter.

### 4.4.5 Operability requirement

Design conditions for operability (serviceability) are defined by the owner. A return period of 1 year is typically specified. Under the defined conditions, the station-keeping system should maintain the vessel within a certain offset limit to allow safe operations of equipment such as processing unit, drilling/production riser, gangway bridge, offloading system, etc. The environmental criteria for operating conditions should be known to the people responsible for the drilling, offloading, or production operations so that operations may be suspended in a timely manner.

### 4.5 Engineering analysis and code check

The main purpose of mooring engineering analysis is to verify that the preliminarily design meets all design requirements in standards, regulations, and project specifications. The intuitively designed mooring system must be verified by engineering analysis which should include the calculation of steady-state loads from wind, waves, and current, the floater's low-frequency and wave-frequency responses, and the predictions of vessel offset and mooring line tensions. The results are checked against the design standards and project specific requirements.

### 4.5.1    Mooring analysis load cases

One important element in the mooring design process is the selection of design load cases. Based on the characteristics of the floater, mooring system, and environmental conditions, a wide range of design load cases should be considered in order to capture the most critical cases. The mooring system is typically designed to withstand the 100-year or even 1000-year return extreme storms. Load cases for maximum design and survival conditions are defined by mooring engineers, taking into consideration all potential combinations of risers and floater loading conditions, and all phases of operation and installation. In some cases, the mooring performance in the operating condition is also analyzed.

The mooring system design for FPSOs should consider floater draft variations. At least two loading conditions, the fully loaded draft and a light ballast draft should be analyzed. In general, the ballast draft is more wind driven while the loaded draft is more wave and current driven. Investigating the two extreme drafts ensures that the most critical condition is captured. Sometimes, it is necessary to investigate an intermediate loading condition for the purpose of extreme and fatigue analysis. Table 4.3 gives an example of a partial list of load cases.

For spread moored systems, the collinear and near-collinear combinations with wind, waves, and current coming from the beam direction are usually the most critical and should be evaluated. For turret mooring systems, noncollinear conditions should be considered. The directional combination of wind, waves, and current can be based on site-specific design data.

According to the rule requirements, the mooring design load cases should include both the mooring intact conditions and one-line damaged conditions. For capturing the maximum tension of the damaged condition, the analysis with the second-most-loaded line removed should be carried out. For capturing the maximum offset in the damaged condition, the analysis with the most-loaded line removed should be carried out. For thruster-assisted moorings, the contribution of thruster output to resist the mean environmental load can be included, and the analysis of damaged mooring condition would include the failure of one thruster.

The engineering analysis will also include the mooring fatigue analysis to verify that the design fatigue lives of all mooring components when factored by the safety factors exceed the field life. For the fatigue analysis, it can be assumed that the floater spends 50% of the time in the loaded draft and 50% of the time in the ballast draft.

**TABLE 4.3** Load cases for mooring analyses.

| Design condition | Environment (load case no.) | Mooring line condition | Safety factor for strength (API RP 2SK) |
|---|---|---|---|
| Maximum design (i.e., extreme) | 100-year (load case 1) | Intact | 1.67 |
| Maximum design (i.e., extreme) | 100-year (load case 2) | Intact | 1.67 |
| Maximum design (i.e., extreme) | – | Intact | 1.67 |
| Maximum design (i.e., extreme) | 100-year (load case N) | Intact | 1.67 |
| Maximum design (i.e., extreme) | 100-year (load case 1) | One line broken | 1.25 |
| Maximum design (i.e., extreme) | 100-year (load case 2) | One line broken | 1.25 |
| Maximum design (i.e., extreme) | – | One line broken | 1.25 |
| Maximum design (i.e., extreme) | 100-year (load case N) | One line broken | 1.25 |
| Survival (i.e., robustness check) | 1000-year or 10,000-year (specified by owner/operator) | Intact | 1.0 |

*API RP*, American Petroleum Institute Recommended Practice.

## 4.6 Questions

1. Designing a mooring system for a water depth of 3000 m, which mooring profile would you choose among chain−polyester−chain, chain−wire−chain, all-chain, or others? Why?
2. What is the issue with a $4 \times 1$ (and $3 \times 1$) mooring pattern? For what kind of field applications, would you consider a $4 \times 1$ or $3 \times 1$ mooring pattern?
3. Name at least one advantage that grouped (clustered) mooring patterns have over evenly spread mooring patterns.
4. Name at least two means in your mooring design to further reduce the vessel offset without adding more lines.
5. Name one mooring design code that is commonly used in the industry. What are the safety factors prescribed for strength in that code?

# References

[1]  K. Ma, A. Ku, C. Chen, T. Kwan, D. Chen, Safety factors in mooring design standards— calibration for consistent reliability, in: Proceedings of the 23rd Offshore Symposium, Society of Naval Architects and Marine Engineers (SNAME), Houston, February 2018.

[2]  BSEE (MMS), Synthetic mooring systems, in: Notice To Lessees NTL No. 2009-G03, 2009.

[3]  H.A. Haslum, J. Tule, M. Huntley, Red Hawk polyester mooring system design and verification, in: OTC 17247, Offshore Technology Conference, Houston, TX, 2–5 May 2005.

[4]  M. Huntley, Polyester mooring rope: length determination and static modulus, IEEE OCEANS 2006, Boston, MA, 2006, pp. 18–21.

[5]  API RP 2SK, Recommended Practice for Design and Analysis of Stationkeeping Systems for Floating Structures, 2005.

[6]  K. Ma, R. Garrity, K. Longridge, H. Shu, A. Yao, T. Kwan, Improving reliability of MODU mooring systems through better design standards and practices, in: OTC 27697, OTC Conference, May 2017.

[7]  ABS, Rules for Building and Classing of Floating Production Installations, American Bureau of Shipping, 2014.

[8]  DNV OS E301, Position Mooring, 2010.

[9]  BV 493-NR, Classification of Mooring Systems for Permanent and Mobile Offshore Units, 2015.

[10]  ISO 19901-7, Petroleum and Natural Gas Industries—Specific Requirements for Offshore Structures—Part 7: Stationkeeping Systems for Floating Offshore Structures and Mobile Offshore Units, 2010.

# Chapter 5

# Mooring analysis

## Chapter Outline

5.1 **Theoretical background** 86
  5.1.1 Governing equations of mooring line 86
  5.1.2 Static solution—catenary equation 87
  5.1.3 Mooring line stiffness 89
  5.1.4 Mooring line dynamics 91
  5.1.5 Mooring system 91
5.2 **System modeling** 93
  5.2.1 Modeling of floaters 93
  5.2.2 Modeling of mooring lines 94
  5.2.3 Modeling of risers 96
  5.2.4 Modeling of environments and seabed 96
  5.2.5 Analysis procedure 96
5.3 **Modeling of polyester rope stiffness** 97
  5.3.1 Upper—lower bound model 98
  5.3.2 Static—dynamic model 99
5.4 **Quasistatic or dynamic analyses** 101
5.5 **Strength analysis in frequency domain** 102

5.5.1 Response transfer functions 102
5.5.2 Frequency-domain analysis procedures 103
5.5.3 Limitation of frequency-domain analysis 104
5.6 **Strength analysis in time-domain** 105
  5.6.1 Time-domain approach 105
  5.6.2 Analysis procedure 106
  5.6.3 Summary 106
5.7 **Uncoupled and coupled analyses** 107
  5.7.1 Uncoupled analysis 107
  5.7.2 Coupled analysis 108
  5.7.3 Industry practice 109
5.8 **Response-based analysis** 109
5.9 **Mooring software** 110
  5.9.1 OrcaFlex by Orcina Ltd. 110
  5.9.2 DeepC/SESAM by DNV GL 110
  5.9.3 Ariane by Bureau Veritas 111
  5.9.4 Other tools 111
5.10 **Questions** 112
**References** 112

The basic principle of mooring system design was introduced in the previous chapter and the engineering analysis to evaluate the mooring system performance is presented in this chapter followed by the fatigue analysis in the next chapter.

As the mooring line catenary configuration is the fundamental element of all mooring analysis, the mooring system theoretical background is first introduced. It is followed by the numerical models to represent the floater, the mooring lines, and risers as building blocks of the overall

Mooring System Engineering for Offshore Structures. DOI: https://doi.org/10.1016/B978-0-12-818551-3.00005-3

station-keeping system. The modern engineering analysis of mooring systems is then explained to predict the floater dynamic responses, mooring line tensions, and riser loads. The commonly used engineering software tools are also described.

## 5.1 Theoretical background

### 5.1.1 Governing equations of mooring line

To understand the primary mechanism of mooring line configuration and tension, let us consider one small element of the mooring line in a 2-D plane (the coordinate system with only $x$ and $z$), as shown in Fig. 5.1. In the free body diagram, let $P$ denote the wet weight per unit length, $T$ the effective tension, $dl$ the length, and $AE$ the axial stiffness. $d\psi(l)$ denotes the displacement in the direction normal to the mooring line, and $d\phi(l)$ the displacement in the tangential direction of the mooring line, as shown in Fig. 5.1.

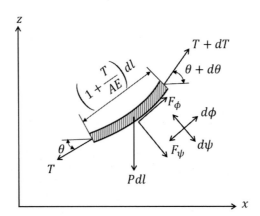

**FIGURE 5.1** Force and displacement on an element of a mooring line.

It is assumed that the line bending and torsional stiffness terms are insignificant and can be neglected. Note this is a reasonable assumption for chain as well as for wire or polyester rope with a large radius of curvature and is well accepted by the industry.

Let $F$ denote the hydrodynamic force acting on the element with a mass $m$, and subscript $\phi$ and $\psi$ the components in element's tangential and normal direction. In $\phi$ and $\psi$ directions, we have the following relationships:

$$-T + (T + dT)\cos d\theta - P\sin\theta\, dl + F_\phi\left(1 + \frac{T}{AE}\right)dl = m\frac{d^2\phi(l)}{dt^2}, \quad (5.1)$$

$$(T + dT)\sin d\theta - P\cos\theta\, dl - F_\psi\left(1 + \frac{T}{AE}\right)dl = m\frac{d^2\psi(l)}{dt^2}. \quad (5.2)$$

For an infinitesimal element $dl$, we have $\cos d\theta = 1$, $\sin d\theta = d\theta$, $dT \, d\theta = 0$. Thus we have the following equations in the tangential and normal directions respectively:

$$dT - P\sin\theta \, dl + F_\phi \left( 1 + \frac{T}{AE} \right) dl = m \frac{d^2\phi(l)}{dt^2}, \tag{5.3}$$

$$T \, d\theta - P\cos\theta \, dl - F_\psi \left( 1 + \frac{T}{AE} \right) dl = m \frac{d^2\psi(l)}{dt^2}. \tag{5.4}$$

In Eqs. (5.3) and (5.4), the hydrodynamic forces $F_\phi$ and $F_\psi$ on a mooring element can be calculated by a number of different methods, including numerical approaches, such as solving the Navier—Stokes equations and obtaining hydrodynamic forces through the integration of the pressure, or experimental methods [1—4]. In most numerical tools for industry $F_\phi$ and $F_\psi$ are calculated through Morison equations for moving structures in waves and currents, which basically are the combination of Morison equations for a fixed structure in moving waters and for a moving structure in still water.

Eqs. (5.3) and (5.4) are based on force dynamic balance. The $(x, z)$ coordinate system has the following relationship with $(l, \theta)$,

$$dx = \left( 1 + \frac{T}{AE} \right) \cos\theta \, dl, \tag{5.5}$$

$$dz = \left( 1 + \frac{T}{AE} \right) \sin\theta \, dl. \tag{5.6}$$

The following relationship can be established from the coordinate rotation between $(x, z)$ and $(\phi, \psi)$:

$$d\psi = dz \cos\theta - dx \sin\theta, \tag{5.7}$$

$$d\phi = dx \cos\theta + dz \sin\theta. \tag{5.8}$$

Eqs. (5.3)—(5.8), along with appropriate boundary conditions on the floating structure's attaching point and seabed conditions, will be the governing differential equations for a mooring line considering both mooring line dynamics and elastics. The equations are nonlinear, and there are no analytical solutions in general. In most cases one has to resort to numerical tools such as finite element method (FEM) to solve them [5—7].

## 5.1.2  Static solution—catenary equation

The static shape of a mooring line hanging under its self-weight from a fairlead to a touch down point at the seabed follows a "catenary" configuration. The shape can be described by the catenary equation that was first derived by Leibniz, Huygens, and Bernoulli in 1691 [8]. The solution of the catenary equation has a concise form of a hyperbolic cosine function.

Let's first ignore the dynamic loads on the mooring line, including damping and inertia forces, as they are small enough so that mooring line geometry and tension distribution are only a function of the top end position. In other words, we are solving the statics of the mooring line, which follows the catenary shape while the floating structure moves on the sea surface.

Assume that the mean horizontal environment force from wind, wave, and current acting on the mooring line at its attachment point is $T_0$. The origin of the reference frame $(x, z)$ is located at the point of contact of the catenary line with the seabed $(x_0, z_0)$, which corresponds to the point of zero slope, as shown in Fig. 5.2. Let $l_S$ denote the total suspended length (arc length), $l_T$ the total mooring line length, and $h$ is the vertical distance from that point to the seabed.

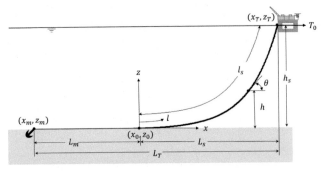

**FIGURE 5.2**   Geometry of catenary line.

For simplicity, let us assume inelastic mooring components, that is, $AE = \infty$. For a catenary mooring line of one single material, Eqs. (5.3) and (5.4) become

$$dT - P\sin\theta \, dl = 0, \tag{5.9}$$

$$T \, d\theta - P\cos\theta \, dl = 0. \tag{5.10}$$

With the boundary conditions at the seabed and at the top connection, Eqs. (5.9) and (5.10) have the solution of the "catenary equation" as follows:

$$l(x) = \frac{T_0}{P}\sinh\left(\frac{P}{T_0}x\right), \tag{5.11}$$

$$h(x) = \frac{T_0}{P}\cosh\left(\frac{P}{T_0}x\right) - \frac{T_0}{P}. \tag{5.12}$$

Based on Eqs. (5.11) and (5.12), one will be able to plot the mooring line profile, given a horizontal tension $T_0$. On the suspended section where $0 < l < l_s$, we have

$$l = \sqrt{h\left(h + 2\frac{T_0}{P}\right)}. \tag{5.13}$$

The tension along this section is given by

$$T(l) = T_0 + P\,h. \tag{5.14}$$

Now let us look at the above catenary equations in a practical context. For a chain-only catenary mooring line, which is still a common choice in shallow water mooring, the above equation means that static chain tension at fairlead increases linearly with environmental force, equating to horizontal force plus the suspended submerged chain weight.

For a mobile offshore drilling unit (MODU) using drag anchors without vertical capacity, it is crucial to estimate the minimum line length to avoid the uplift force at the anchor, which can be derived by the following relationship between catenary length $l$ and total tension $T$:

$$l = h\sqrt{\left(2\frac{T}{P\,h} - 1\right)}. \tag{5.15}$$

To illustrate by an example, let us assume a 5-in. (5″) R4 studlink chain system, with the chain wet weight $P = 3.0$ kN/m at a water depth of 50 m. The minimum breaking strength (MBS) for such a chain is 14,955 kN, and pretension is set to be 20% of MBS, that is, $T = 2991$ kN. To avoid vertical load on the drag anchor, the minimum line length from the fairlead to touch down point will be $l = 312$ m based on Eq. (5.15).

The solutions in Eqs. (5.11) and (5.12) are for a single-material inelastic catenary mooring line. It is possible to get quasistatic solutions for multimaterial catenary lines considering elastic elongation [9–11].

### 5.1.3   Mooring line stiffness

A mooring line as depicted in Fig. 5.2 will exert a horizontal and vertical force, $T_H$ and $T_v$ respectively, on the floating structure. These two forces $T_H$ and $T_v$ have a relationship with the floating structure's offset. The larger the floating structure's offset from its equilibrium position, the larger the reaction force $T_H$ will be. Similar to a simple spring system, we call such a relationship between floating structure offset and mooring line reaction force as mooring line stiffness. For the same reason, sometimes we call the reaction force as the restoring force.

Mooring stiffness gives a proportional relationship between force and displacement. When the line top tension increases, the mooring line will have axial elongation as well as overall geometric deformation. Therefore mooring stiffness comprises of stiffness contribution from axial stiffness $AE$ as well as geometric stiffness, as illustrated in Fig. 5.3 for a quasistatic analysis of a mooring line.

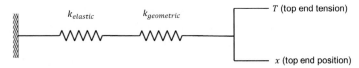

**FIGURE 5.3**   Illustration of a quasistatic analysis of a catenary mooring line.

The total stiffness of a catenary line is the combination of elastic and geometric stiffness:

$$\frac{1}{k_{Total}} = \frac{1}{k_{elastic}} + \frac{1}{k_{geometric}}. \tag{5.16}$$

The *elastic stiffness* contribution to mooring line stiffness $k_{elastic}$ can be defined as a function of $AE$ divided by total line length $l$, that is, $AE/l$.

To understand the *geometric stiffness*, let us assume an inelastic mooring line, that is, $k_{elastic} = +\infty$, and look at line stiffness resulting from geometric stiffness. Assume the floating structure has a very small oscillation around the equilibrium position. Denoting $\eta_1$ and $\eta_3$ the horizontal and vertical motions in the $x$ and $z$ directions, respectively, we have:

$$T_H = (T_H)_M + k_{11}\eta_1, \tag{5.17}$$

$$T_v = (T_v)_M + k_{33}\eta_3. \tag{5.18}$$

In the above equations, $k_{11}$ and $k_{33}$ are stiffnesses in the horizontal and vertical directions, respectively. To obtain explicit expression of $k_{11}$ and $k_{33}$, we need the relationship between a small change in $dT$ resulting from small displacement change $d\eta$. From the geometric relationship as shown in Fig. 5.2, we have $L_T = l - l_s + x$. Combining it with Eqs. (5.12) and (5.13) gives

$$L_T = l - \sqrt{h\left(h + 2\frac{T_H}{P}\right)} + \frac{T_H}{P}\operatorname{arccosh}\left(\frac{Ph}{T_H} + 1\right). \tag{5.19}$$

Differentiating the above equation with respect to $T_H$, we obtain the following analytical expression of mooring line horizontal stiffness due to line overall geometric deformation:

$$k_{11} = \frac{P}{\left(\operatorname{arccosh}\left(\frac{Ph}{T_H} + 1\right)\right) - \left(2/\left(\sqrt{1 + \left(2\frac{T_H}{Ph}\right)}\right)\right)}. \tag{5.20}$$

In the vertical direction, from Eq. (5.13) we have: $T_v(l) = Pl = P\sqrt{h^2 + 2(T_0/P)h}$, and thus differentiating this equation with respect to $h$ we have the vertical geometric stiffness:

$$k_{33} = \frac{Ph + T_0}{\sqrt{h^2 + (2(T_0/P)h)}} \tag{5.21}$$

### 5.1.4    Mooring line dynamics

The following equation describes the mooring line tension caused by the wave frequency (WF) motion:

$$M\frac{d^2r}{dt^2} + B\frac{dr}{dt} + K\,r = F(r,t), \tag{5.22}$$

where $M$ is mass including added mass, $B$ is damping, $K$ is stiffness matrix, $F$ is an external exciting force, and $r = (x, y, z)$ is displacement vector from the mean position. The dynamic analysis, including added mass, damping, and stiffness are illustrated in Fig. 5.4.

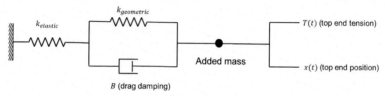

FIGURE 5.4    Illustration of a dynamic analysis for a catenary mooring line.

The finite element (FE) method is a popular approach to solve dynamic line equations [6,7,12].

### 5.1.5    Mooring system

Let us consider a mooring system with multiple lines connected with an offshore floating structure, as shown in Fig. 5.5.

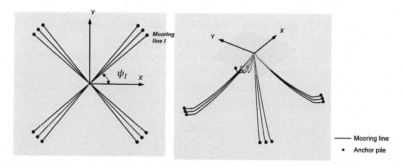

FIGURE 5.5    Mooring system with multiple mooring lines.

The floating structure motion can be expressed as the equation of motions in six degrees of freedom:

$$\sum_{j=1}^{6}\left[\left(M_{ij} + M_{aij}\right)\frac{d^2\eta_j}{dt^2} + B_{Lij}\frac{d\eta_j}{dt} + B_{Qij}\left|\frac{d\eta_j}{dt}\right|\frac{d\eta_j}{dt} + K_{ij}\eta_j\right] = F_i \tag{5.23}$$

where indices $i$ and $j$ indicate the direction of the fluid force and the mode of motion respectively. $i = 1,2,3,4,5,6$, refers to surge, sway, heave, roll, pitch, and yaw, respectively. $B_L$ and $B_Q$ are linear and quadratic damping coefficients, respectively. The right-hand side of Eq. (5.32) shall include environmental forces, including mean environmental force from wind, wave, current, and WF and low-frequency (LF) force from wave and wind as well as wave−current interaction, as discussed in Chapter 3, Environmental loads and vessel motions. For floating structures with risers, umbilicals, and flowlines, the static and dynamic pulling force shall also be accounted for in Eq. (5.23).

For quadratic damping terms, sometimes it is possible to linearize the nonlinear quadratic term and write the two damping terms in the form of

$$\left(B_L + B_Q \left|\frac{d\eta}{dt}\right|\right)\frac{d\eta}{dt} = B\frac{d\eta}{dt}. \tag{5.24}$$

The linearization of drag quadratic damping terms is necessary for a frequency-domain analysis as detailed in Refs. [13,14].

Not all damping is a result of the viscous effect from fluids. There are other sources of damping, such as potential damping and structural damping. The potential damping is similar to the added mass resulting from the radiating waves emanating from the body when it is forced to oscillate.

For large offshore structures such as Semi, Spar, or floating production storage offloading (FPSO) units, motion damping is due to wave radiation damping, hull skin friction damping, eddy making damping, viscous damping from bilge keels and other appendages, and viscous damping from risers and mooring lines.

On the left-hand side of Eq. (5.23), the added mass, damping, and stiffness from mooring lines should be taken into account. For the stiffness matrix $K$ and damping matrix $B$, the contribution from mooring lines is the summation of the contributions of all the mooring lines under the six degrees of motions. For the surge and sway directions, the total stiffness coefficients from moorings as shown in Fig. 5.5 are:

$$\text{Surge: } K_{11} = \sum_{I=1}^{n} k_I\cos^2\psi_I, \tag{5.25}$$

$$\text{Sway: } K_{22} = \sum_{I=1}^{n} k_I\sin^2\psi_I. \tag{5.26}$$

Natural frequencies (periods) and critical damping are essential properties for a floater's motion. A resonance motion occurs significantly when the external excitation force period is near the natural period of a floating system. In a mooring analysis, we might need to calibrate the total damping level as a ratio to critical damping with model test results using a free decay test. The natural frequency $f$ and critical damping $B_C$ of a floater system are defined as functions of the total mass $(M + M_a)$ and stiffness $K$ in the six degrees of freedom:

$$\text{Natural frequency:} f = \sqrt{\frac{K}{M + M_a}}, \qquad (5.27)$$

$$\text{Critical damping:} B_C = 2\sqrt{(M + M_a)K}. \qquad (5.28)$$

## 5.2    System modeling

### 5.2.1    Modeling of floaters

To perform mooring dynamic analysis would require the inputs of the floater such as the mass properties, the center of gravity, mass moment of inertia, and radii of gyration, etc. The mooring analysis would also require the floater's hydrodynamic coefficients and motion characteristics. For large floating structures such as a semisubmersible or FPSO, mean wave drift coefficients, added mass, potential damping, and motion response amplitude operators (RAOs) can be obtained from a radiation/diffraction analysis. The details of floater loads and motions are described in Chapter 3, Environmental loads and vessel motions, of this book.

Radiation/diffraction programs are based on nonviscous flows, and thus viscous drift force is not captured in frequency-domain analysis. Therefore the correction of viscous drift can be added to the results from radiation/diffraction analysis. In the time-domain analysis, one can estimate the viscous drift force from the "Morison model." The Morison model includes Morison elements representing the hull components. The Morison elements should be modeled high enough above the mean waterline to capture the actual wave elevations.

Motion RAOs may be used directly for WF motion calculation in a frequency-domain analysis. Other than applying motion RAOs, a time-domain analysis can use force RAOs in a combination of added mass and retardation functions. A retardation function represents the memory effects of fluid and can be obtained from the added mass and damping terms.

## 5.2.2 Modeling of mooring lines

The line modeling defines the properties of mooring lines, including the geometric shape, weight, and axial stiffness. With defined fairlead coordinates, anchor coordinates, line segment lengths, unit weight and stiffness of each segment, the line catenary shape is determined. Alternatively, the analyst can provide the pretension and have the software to calculate the anchor radius (the horizontal distance from fairlead to anchor).

Actually for a given water depth without any environmental loads, the catenary of the mooring line has a relation among the following three variables. With any two of the three defined, the other can be calculated.

- *Anchor radius* (i.e., the horizontal distance from the fairlead to the anchor).
- *Length of the line* (i.e., the total length composed by segments of certain weights).
- *Pretension at the top* (i.e., the tension at fairlead under zero environment).

The manufacturers of the mooring components usually provide the properties of mooring lines. For the mooring analysis, the MBS, wet weight, and axial stiffness are the most fundamental properties of mooring components. If no manufacturer data are available, the formulas provided in Table 5.1 can be applied in the modeling [7,15−18].

**TABLE 5.1** Breaking strength, unit weight, and axial stiffness of mooring components.

| Mooring component | | MBS (kN) | Submerged weight (N/m) | Axial stiffness (kN) |
|---|---|---|---|---|
| **Chain** | R3 | $0.0223 \times d^2 (44 - 0.08d)$ | Stud: $0.187 \times d^2$ Studless: $0.171 \times d^2$ | Stud: $101 \times d^2$ Studless: $85.4 \times d^2$ |
| | R3S | $0.0249 \times d^2 (44 - 0.08d)$ | | |
| | R4 | $0.0274 \times d^2 (44 - 0.08d)$ | | |
| | R4S | $0.0304 \times d^2 (44 - 0.08d)$ | | |
| | R5 | $0.0320 \times d^2 (44 - 0.08d)$ | | |
| **Wire rope (spiral strand)** | | $0.9 \times d^2$ | $0.043 \times d^2$ | $88.7 \times d^2$ |
| **Polyester rope** | | $0.25 \times d^2$ | $0.0017 \times d^2$ | $1.1 \times d^2$ |

*MBS*, Minimum breaking strength. Note: *d* is the mooring component (chain, wire rope, or fiber rope) nominal diameter in mm. Refer to Section 5.3 for the detailed modeling of polyester rope stiffness.

It can be seen that with the same MBS wire rope is lighter than chain. The polyester rope has a much lighter unit weight even than wire rope and is almost neutral in water. The polyester rope also has a much lower axial stiffness than steel wire ropes and chains, and will be further discussed in the next section.

For a permanent mooring system, allowances for corrosion and wear of a mooring chain or unsheathed wire rope are to be included in the modeling. For mooring chains, a corrosion and wear allowance is provided by an appropriate increase in the chain diameter. If site-specific corrosion data are not available, the corrosion allowance recommended in the industry standards can be applied [19−21]. For conservatism, it is a common practice to use the MBS of the corroded chain for calculating the safety factors, while the uncorroded chain is used to model the line properties including weight, drag diameter, and elastic modulus.

For permanent mooring systems, modeling also needs to consider marine growth. The thickness and specific density of marine growth are dependent on the site and are usually provided as a part of the metocean data in the design basis. If no data are available, the recommended value provided by Class or other codes may be used [19,21]. The analysis is typically conducted both with and without marine growth.

The drag coefficient $C_d$ and added mass coefficient $C_m$ of mooring components depend on parameters such as Reynolds number, Keulegan−Carpenter number, and surface roughness number. These coefficients are empirically determined in some Class Rules [22]. Table 5.2 lists typical values of hydrodynamic coefficients for different mooring and riser components. The corresponding line segment nominal diameter $d$ shall be used for drag and added mass forces estimation.

**TABLE 5.2** Examples of mooring component hydrodynamic coefficients.

| Line segment | $C_d$ | | $C_m$ | |
|---|---|---|---|---|
| | Normal | Tangential | Normal | Tangential |
| Studless chain | 2.4 | 1.15 | 2.0 | 1.0 |
| Stud chain | 2.6 | 1.4 | 2.0 | 1.0 |
| Polyester rope | 1.6 | 0.1 | 1.0 | 0.1 |
| Sheathed wire rope | 1.0 | 0.1 | 1.0 | 0.1 |
| Bare riser/umbilical | 1.0 | 0.1 | 1.0 | 0.1 |

### 5.2.3   Modeling of risers

Risers, umbilicals, and flowlines, if any, will introduce additional stiffness, loads (lateral pull), damping, and inertia effects on the floater. Some of the floating production units are equipped with steel catenary risers or midwater flowlines arranged in asymmetric patterns, which may impose large riser or flowline loads on the mooring system. In such a case, the impact of riser or flowline loads should be included.

A floating production unit might have some initial risers and require potential tie-in of future lines during the service life. All possible riser scenarios, based on the anticipated installation sequence and future tie-in of additional risers, need to be considered in the riser modeling.

### 5.2.4   Modeling of environments and seabed

Excitation forces from wind, wave, and current, as discussed in Chapter 3, Environmental loads and vessel motions, need to be entered in the analysis. Most advanced mooring software uses a wind spectrum to capture dynamic wind effects. Waves are usually modeled as wave spectra. The current velocity is used to calculate the steady current forces. For time-domain simulations, where the current force on the hull is modeled with Morison elements, adjustments of the current speed or drag coefficients might be required to account for shielding effects.

If the seabed is relatively flat, it is generally acceptable to model the seabed as a horizontal flat surface. In situations where the seabed exhibits a marked slope or change in shape within the scope of the mooring system, the seabed bathymetry needs to be taken into consideration.

### 5.2.5   Analysis procedure

A typical mooring analysis would include the following tasks:

1. Computation of environmental loads.
2. Floater hydrodynamic analysis to determine the hydrodynamic coefficients.
3. Mooring static analysis.
4. System LF motion analysis.
5. System dynamic analysis to determine the floater response and line tensions.

Fig. 5.6 illustrates a typical mooring system design and analysis procedure. The mooring analysis can be performed either with a frequency-domain method or with a time-domain method. The procedure shows a frequency-domain approach, which requires motion RAO as an input file.

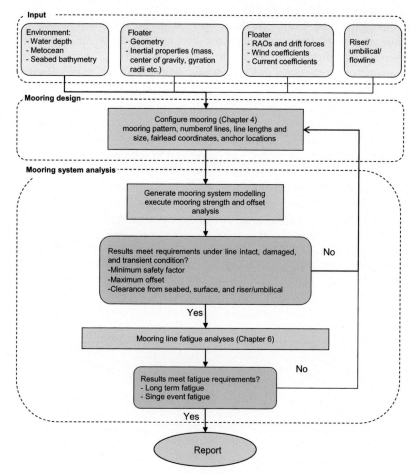

**FIGURE 5.6**   Example of mooring analyses workflow.

## 5.3   Modeling of polyester rope stiffness

Polyester ropes are made of materials with viscoelastic properties, so their stiffness is not constant. It varies with the load duration and magnitude, the number and frequency of load cycles, and the loading history. Also, polyester mooring lines become stiffer after a long time in service. Historical loading above a certain level may lead to a permanent increase of the rope length and results in a softer mooring system, if no retensioning (i.e., shortening) is performed.

The rope also behaves in two fundamentally different modes of stiffness. When the rope is loaded rapidly, its stiffness becomes higher, which is defined as dynamic stiffness. When the rope is loaded slowly, the stiffness is lower, which is defined as static stiffness. Because of this rope behavior, it is not possible to develop models that represent the precise stiffness characteristics of the rope. Currently, the industry relies on simplified models that capture the most important characteristics and at the same time yield conservative predictions of line tensions and vessel offsets. Two models are commonly used: the *upper−lower bound model* and the *static−dynamic model* [23−25].

### 5.3.1  Upper−lower bound model

The upper−lower bound model was introduced in response to the need from the industry for a practical stiffness model for polyester mooring systems [19,23]. This model defines lower bound (postinstallation) and upper bound (storm) stiffness values as an approximation. The lower bound represents the static rope stiffness, and the upper bound the dynamic stiffness. These lower and upper bound values are then used to calculate maximum offsets and line tensions, respectively. A plot of typical upper bound stiffness (at $30 \times$ MBS) and the lower bound stiffness (at $10 \times$ MBS) is shown in Fig. 5.7 as an example. Note that the approximation based on the rope diameter, introduced in Section 5.2, is rarely used in a detailed engineering.

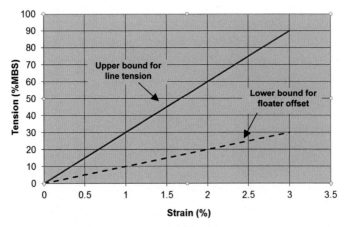

**FIGURE 5.7**  Upper−lower bound stiffness model for polyester ropes.

The model has been widely used in the industry due to its simplicity. However, it does have certain shortcomings. First, there is no systematic method to determine the upper and lower bound stiffness, and therefore these values are often empirically based. Second, the polyester rope has a complicated stiffness property, which is a function of load duration, magnitude, amplitude, and history. Using two limiting values to represent the complicated behavior may result in overly conservative or nonconservative analysis results.

For the upper−lower bound model, the mooring analysis is performed twice, one with the lower bound stiffness and another one with the upper bound stiffness. It is important to ensure the pretension for both runs is the same. Then the maximum offset is determined by the first run, and the maximum line tension is determined by the second run [23].

### 5.3.2 Static−dynamic model

The static−dynamic model was developed to account for the behavior that the dynamic stiffness of polyester ropes can be two to three times the static stiffness [26]. In this model, two slopes are used instead of one constant stiffness (see Fig. 5.8). The static stiffness is utilized for the initial region of the loading curve up to the mean load. Afterward, the dynamic stiffness is used to predict the cyclic part of the loading including low and wave frequencies [23]. This model more accurately simulates the actual conditions faced by a polyester mooring. A mooring line under a severe environment typically experiences a steady mean load and dynamic loads oscillating around the mean load. Typical static stiffness is in the range of 10−15 times MBS, and typical dynamic stiffness is in the range of 25−35 times MBS.

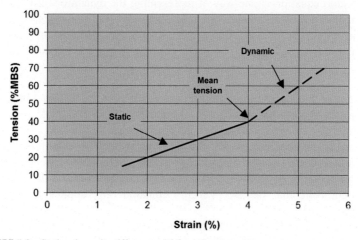

**FIGURE 5.8** Static−dynamic stiffness model for polyester ropes.

Additionally, if the rope is held at a static load, it will continue to elongate or creep. Taking into account this additional elongation at a static load, a so-called quasistatic stiffness, which is lower than the elastic static stiffness, may be defined to serve as a better model than the static stiffness [23].

Static (or quasistatic) stiffness is a function of installation load, static load level and duration, age of rope, and creep property, which can also be determined by testing [23].

Table 5.3 shows an example of polyester rope stiffness values for different purposes. These values may be used for preliminary analyses, before the stiffness values based on production rope tests become available.

**TABLE 5.3** Example of polyester rope stiffness value sets for different analysis purposes.

| Purposes | Static stiffness | Dynamic stiffness (low and wave frequency) |
| --- | --- | --- |
| For calculating vessel offset | 10 × MBS | 20 × MBS |
| For calculating line tension | 20 × MBS | 30 × MBS |
| For calculating fatigue damage | | |

MBS, Minimum breaking strength.

Ideally, the polyester rope's load—elongation properties should be modeled as nonlinear elastic by expressing the load—elongation relationship. The Del Vecchio equation is one way of representing the dynamic stiffness. Note that it makes a mooring analysis complex, however. For polyester ropes, Del Vecchio expressed the dynamic stiffness (specific modulus) of polyester ropes as follows [24,27]:

$$SM = C + \alpha L_m + \beta T + \gamma \log P,$$

where $SM$ is the specific modulus defined as stiffness divided by MBS, $C$ is the constant, $L_m$ is the mean load as percentage of MBS, $T$ denotes load amplitude as percentage of MBS, and $P$ is the loading period in seconds. The dimensional coefficients $\alpha$, $\beta$, $\gamma$ are mean load, load amplitude, and load period coefficients, respectively. These coefficients are determined through tests on ropes and will depend on rope construction and the polyester yarn used. Note that the Del Vecchio equation only applies to dynamic stiffness.

For the static—dynamic model, the mooring analysis is also performed twice, once with the static stiffness and again with the dynamic stiffness. It is important to ensure the pretension for both runs is the same. Then the mean responses (tension and offset) from the first run are combined with the dynamic responses from the second run to yield the final results [23].

## 5.4 Quasistatic or dynamic analyses

The main difference between the quasistatic and dynamic analyses is how to handle the WF responses of the floater. In the dynamic analysis, the time-varying fairlead motions are calculated [28] and input to the mooring line dynamic analysis, capturing the time-varying effects due to mooring line added mass, drag forces, damping, and acceleration relative to the fluid. On the contrary, a quasistatic analysis ignores the inertial and damping forces acting on the line. The shape of the mooring line and the tension distribution along the mooring line are functions of the top end positions only. Fig. 5.9 illustrates the difference between the quasistatic and dynamic analysis of a mooring line.

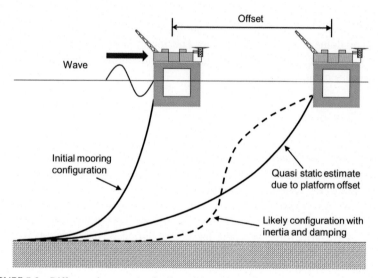

**FIGURE 5.9** Difference between quasistatic and dynamic analysis.

Quasistatic analysis has the advantage of simplicity. However, the quasistatic analysis is appropriate for calculating the mooring line response due to the mean offset and the LF floater motions, and the dynamic analysis is required to predict the mooring line responses due to WF floater motions. Kwan and Bruen [28] have demonstrated that the ratio between dynamic and quasistatic tensions can be significant after studying a total of 13 floaters, including semisubmersibles and ship-shaped vessels in the Gulf of Mexico and the North Sea.

In general, it is recommended to perform a dynamic analysis for both permanent and MODU mooring systems. If WF impact is negligible, the quasistatic analyses can still be used on mooring systems, such as those installed in protected waters. With advanced computing capability, the time saving from running a quasistatic analysis is negligible.

## 5.5 Strength analysis in frequency domain

The frequency-domain analysis is used because of its efficiency. It is extensively used, for example, to design MODU moorings, as a quick solution may be needed during a rig move. It is also often used for calculating fatigue damage where a large number of load cases need to be evaluated.

The frequency-domain analysis assumes that the system response can be expressed as static and frequency-dependent components such that the principle of superposition over different frequencies can be applied. The floater's motions are divided into three parts: mean, WF, and LF components. The same holds for the corresponding mooring line tensions. In the frequency-domain analysis, the floater and the mooring system are typically de-coupled, which means that floater's motions and mooring line tensions are solved separately.

### 5.5.1 Response transfer functions

Both WF and LF mooring analysis involve a vital concept called the response transfer function. A response transfer function builds a bridge connecting incident wave spectrum in WF analysis or wind spectrum in LF calculation to the response spectrum of our interest.

To look at this concept in more detail, let us denote $H(\omega)$ as the response transfer function for motion $\eta$. Here $\eta$ can also be replaced by any other term such as wave drift force once the basic assumption of linearization is satisfied such that the response can be superimposed with different frequency components. Physically, $H(\omega_i)$ is the response per unit wave amplitude at WF $\omega_i$. For example, for an incident wave with amplitude $a_i$, the response of motion $\eta_j$ will be $H(\omega)a_i$. In reality, there might be a phase difference between incident wave and response, but we are neglecting it here for simplicity.

Recalling Chapter 3, Environmental loads and vessel motions, the wave spectrum is proportional to the square of wave amplitude. We therefore have the following relationship between wave amplitude $a_i$ and wave spectrum $S_w(\omega_i)$:

$$\frac{1}{2}a_i^2 = S_w(\omega_i)\Delta\omega, \tag{5.29}$$

where $a_i$ and $\Delta\omega$ denote wave amplitude and constant difference between successive frequencies, respectively. Since we are assuming that all responses in different frequencies can be linearly superimposed, the response spectrum for motion $\eta$ will be:

$$S_\eta(\omega) = |H(\omega)|^2 S_w(\omega). \tag{5.30}$$

With the motion response spectrum defined, it is now straightforward to get the standard deviation and the statistical parameters, such as significant value, most probable maximum value, and expected extreme value. In frequency-domain analysis, the critical step is to obtain the response transfer function $H(\omega)$.

For WF motions, the transfer functions (RAOs) are calculated from radiation-diffraction theory. For LF motions, the motion is computed by solving the *linearized* equation of motion oscillating around the mean position $\bar{\eta}$, that is,

$$M\frac{d^2\eta}{dt^2} + B\frac{d\eta}{dt} + C\eta = F^{LF}, \tag{5.31}$$

where $M$, $B$, $C$, and $F^{LF}$ denote linear effective mass, damping, stiffness, and LF load, respectively. Since the above equation is a linear equation, and $\eta$ can be expressed as sine or cosine wave functions, with some mathematical transformation we can have the following transfer function for LF motion:

$$H^{LF}(\omega) = \left(-\omega^2 M + j\omega B + C\right)^{-1}. \tag{5.32}$$

### 5.5.2   Frequency-domain analysis procedures

API RP 2SK provides a frequency-domain analysis procedure for spread mooring and single-point mooring systems [19]. The general equations of motion describing the response of the structure are analyzed separately for mean, low, and WF responses.

1. Define the wind, wave, and current conditions for the load case being analyzed.
2. Apply the mean environmental loads on the moored floater to determine the equilibrium mean position of the floater.
3. Determine floater's LF motion based on the mooring stiffness at the equilibrium position.
4. Determine floater's WF motion using the RAOs obtained from the hydrodynamic motion analysis.
5. Mooring line tensions and vessel offset are calculated by the dynamic analysis based on the predicted floater's motions.
6. Check if the line tensions and vessel offset meet the design criteria.

Physically, the LF and WF components, either for motion or tension, are interacting with each other. Since they are analyzed separately, the combination of these two components is difficult. In the current industry practice, it is common to calculate extreme offset and tension during a storm or hurricane based on API RP 2SK [19]. The maximum values of the offset $\eta_{max}$ can be determined by:

$$\eta_{max} = \eta_{mean} + \max\left(\eta_{LF\ max} + \eta_{WF\ Sig}, \eta_{LF\ Sig} + \eta_{WF\ max}\right),$$

where subscript LF and WF mean components of low frequency and wave frequency, and subscripts Sig and max mean significant value and maximum value. The maximum tension $T_{max}$ can be determined by:

$$T_{max} = T_{mean} + \max\left(T_{LF\ max} + T_{WF\ Sig}, T_{LF\ Sig} + T_{WF\ max}\right).$$

This method is considered conservative by combining the alternating maximum and significant values of the low and WF components.

Particular attention should be given to the LF damping of the mooring system. LF motion is narrow-banded since it is dominated by the resonant response at the natural frequency of the moored structure. Therefore the motion amplitude is highly dependent on the stiffness of the mooring system and the damping. There are three primary sources of LF damping: viscous damping of the structure; wave drift damping of the structure; and mooring (and riser if applicable) system damping.

If no damping data is available from direct calculation or model test, the empirical formula recommended by Bureau Veritas (BV) can be used as extra linear damping coefficients in the surge, sway, and yaw directions for semisubmersibles or ship-shaped tanks [29]. For a semisubmersible unit at operating draft, the linear damping coefficients are about 10% of critical dampings in the surge and sway directions and 5% in the yaw direction. For ship-shaped floaters, the linear damping coefficient varies depending on the mooring line patterns and hull shapes.

### 5.5.3 Limitation of frequency-domain analysis

The frequency-domain analysis approach has the following limitations:

1. Equations of motions are linearized. Caution is required when using the frequency-domain analysis for shallow water mooring, mooring line with clump weight or buoy, and any other scenarios in which nonlinearity might be significant. Also linearization is response amplitude dependent.
2. The actual extreme value distribution is not obtained directly from the analyses, and the extreme values are derived based on an assumed statistical distribution.
3. The combination of the WF and LF components is empirical.

For a floater with a single-point mooring system that allows weather-vaning, the time-domain simulation or model testing may be most appropriate, as the floater may experience large LF yaw motions. In order to perform a mooring analysis based on the approach of the frequency domain, one must make a particular assumption on floater's heading. A common approach is to use the design heading, at which the mooring system responses are calculated, as the equilibrium heading of the floater under mean environmental loads plus or minus the significant LF yaw motions.

## 5.6    Strength analysis in time-domain

### 5.6.1    Time-domain approach

Time domain analysis solves the response of floater and mooring in the time domain through numerical integration. It can model the following nonlinearities:

1. Geometric nonlinearity—associated with large changes in the shape of the mooring line.
2. Sea bottom nonlinearity—the interaction between the mooring line and the seafloor.
3. Direct fluid loading—proportional to the square of relative velocity.

It is therefore considered more time-consuming but more accurate compared with the frequency-domain analysis.

Also, the time-domain method may be used to perform the coupled simulation of mean, low, and WF responses of the vessel and the mooring/riser system. This coupled approach requires a time-domain mooring analysis, which solves the equations of motion for the responses of the vessel, mooring lines, and risers simultaneously.

In the time-domain analysis, the system dynamic equations are solved numerically in which the mooring line is discretized into the number of elements. Most commercial software tools use a FE method [30] in which the line is either simplified as slender members or lumped mass [7,31]. The lumped mass method is one category of the FE methods. A lumped mass method is distinguished from a typical FE method in that in a lumped mass approach the line is divided into a number of nodes that are treated as point masses and springs, while in a typical FE method, mass is distributed along the entire element, leading to a consistent mass formulation. Besides the FE method, there are other methods including the finite difference method, which discretizes over both space and time. Although there are various algorithms, most models achieve similar results as long as a sufficiently finite discretization is used.

## 5.6.2   Analysis procedure

The procedure of the time-domain analysis recommended by API is summarized as follows [19]:

1. Define the wind, wave, and current conditions for the load case being analyzed.
2. Determine the vessel's wind and current force coefficients, and hydrodynamic model of the system including vessel, riser, and mooring.
3. Perform a time-domain simulation for the storm duration using a time-domain mooring software. Repeat the simulation several times using different seeds for generating the wave and wind time histories.
4. Use statistical analysis techniques to establish the expected extreme values of mooring line tension and vessel offset.
5. Check if the line tensions and vessel offset meet the design criteria.

The required length of the simulation depends on many factors, such as wave periods and LF responses, the degree of nonlinearity, and system damping. The 3-hour duration is generally sufficient for the standard deviation of WF responses because it represents about 1000 cycles with a period of 10 seconds. However, LF responses for deepwater systems typically have periods of several minutes. A 3-hour simulation may contain fewer than 50 cycles, which may be insufficient to provide a good statistical confidence. Therefore either simulation of longer duration or repetition of the simulations would be required.

The extreme value of a particular response parameter (vessel offset, line tension, etc.) in a single time-domain simulation will vary. Consequently, repetition of the simulation is required to establish reasonable confidence in the predicted extreme response.

In summary, the time-domain analysis involves statistical processing of simulated time histories to predict the extreme values of vessel offset, line tension, anchor load, or ground line length.

## 5.6.3   Summary

The advantage of the time-domain analysis is its capability to model all system nonlinearities, including mass, damping, and stiffness terms, and time-varying load as input. However, the computation can be time-consuming. Nevertheless, with the advance in computing technology, the time-domain analysis is getting more and more popular and represents the way forward. It is especially valuable when used for analyzing the critical load cases, verification of the frequency-domain analysis, and systems with high nonlinearities, etc. [19].

## 5.7 Uncoupled and coupled analyses

Traditionally, the floater motion analysis and mooring analysis are uncoupled and conducted separately. This means that output from the motion analysis is used as an input for the mooring analysis, while the mooring lines and risers are often simplified in the motion analysis to minimize the computational efforts. However, in reality, the floater and mooring/riser lines are physically coupled with each other. It is particularly true when the number of mooring/ riser lines is large or the floater (vessel) is relatively small. Therefore the coupled analysis provides physically correct modeling of the whole system.

### 5.7.1 Uncoupled analysis

As illustrated in Fig. 5.10, the uncoupled analysis consists of the following two steps:

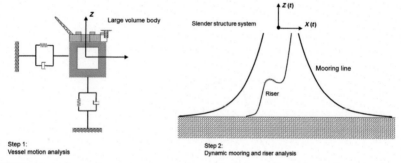

**FIGURE 5.10** Uncoupled analysis: vessel motion and mooring analyses are performed separately.

*Step 1 Calculation of floater motion*—Floater offset and motions are computed in which the mooring lines are modeled as nonlinear position-dependent forces (stiffness). In other words, feedback from mooring dynamic loads to floater motions is not included.

*Step 2 Dynamic analysis of the mooring system*—The top end of the mooring line is exposed to the forced displacement based on the floater motions.

Uncoupled analysis leads to two main simplifications when the WF and LF motions of the floating structure are analyzed:

1. The floater motion is assumed to be not affected by the mooring system dynamics.
2. The damping forces from the mooring lines and risers are neglected or linearized. The damping forces are important for accurately predict the LF motion.
3. Direct current forces on the mooring and riser system are neglected.

The shortcomings of the uncoupled analysis are especially apparent in deep water with strong current and with a large number of moorings and risers. In such a scenario, the mooring system dynamics are significant, the damping due to the mooring and risers is large, and the direct current force on moorings and risers is significant. The uncoupled analysis may fail to predict the system response accurately.

## 5.7.2 Coupled analysis

In the coupled analysis, the floater dynamic responses together with its moorings and risers are solved simultaneously as illustrated in Fig. 5.11 and thus all dynamic interactions are captured [32,33]. Specifically the coupling effects from mooring/riser forces, damping, and inertia forces are governed by the following contributions [21]:

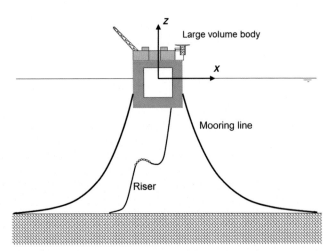

**FIGURE 5.11** Coupled analysis: the dynamics of vessel and mooring lines are analyzed simultaneously.

1. *Restoring*—Static restoring force from the mooring and riser system as a function of floater offset.
2. *Restoring*—Current loading and its effects on the restoring force of the mooring and riser system.
3. *Restoring*—Seafloor friction (if mooring lines and/or risers have bottom contact).
4. *Damping*—Damping from mooring and riser system due their motions (transverse in particular) and, current, etc.
5. *Damping*—Friction forces due to hull/riser contact.
6. *Inertia*—Additional inertia forces due to the mooring and riser system.

The time-domain analysis is used to represent the coupled floater/mooring/riser response at every time instant. In this way, the full interaction between mooring lines, risers, and the floater is taken into account, and accurate floater motions and dynamic loads in mooring lines and risers could be obtained simultaneously. There are many coupled analysis developments, and the detailed implementations of the numerical schemes can be found in the literature [32−36].

### 5.7.3    Industry practice

The uncoupled approach is widely used in the offshore industry and is a valuable tool if properly calibrated and validated. Uncoupled modeling with empirical adjustments, based on model tests and project experience, will remain a practical tool, especially in a project's early engineering phase. Coupled modeling is physically more correct with the improved accuracy. It is commonly applied for the verification of uncoupled analysis, especially for governing design load cases.

API recommends a coupled analysis if there are strong interactions between the floater and mooring/risers [19,20]. The major disadvantage of coupled analysis is that it could be very time-consuming. With advances in computing technology, the application of the coupled analysis is gradually expanding.

The significance of the coupled analysis is also due to the increased offshore activity in deep and ultradeep waters where the coupling between the floater and moorings/risers is more significant, and also model tests with complete mooring systems for design verification are more difficult. The coupled simulation can be utilized together with model tests to form the so-called hybrid model test method as detailed in Chapter 7, Model tests.

## 5.8    Response-based analysis

The traditional approach for the design and analysis of mooring systems is to use a specific $N$-year (e.g., 100-year) return period metocean event and obtain the extreme response. This method is called the design environmental condition approach. The approach is based on the assumption that the $N$-year extreme response occurs precisely during $N$-year environment conditions.

However, a floating structure with a mooring system is likely to undergo dynamic response that depends not only on the dominating environmental condition (e.g., wind speed) but also on other factors such as wave height, wave period, and directions. For this reason, the $N$-year environmental conditions may not always result in the $N$-year "maximum" response. Therefore the traditional approach might underestimate the response.

The response-based analysis (RBA) uses the real long-term database of metocean events including wind, wave, and current to calculate response statistics directly and predict *N*-year extreme response. Thus, in theory, RBA provides a more accurate prediction of responses.

The RBA techniques have been used in structural design for many years to explore the probability of failure of an offshore structure [37,38]. For the mooring system analysis, the RBA can provide a better understanding of the risk originated from the response [39,40]. Due to a large number of load cases or sea states to handle, the RBA will be more time-consuming than the conventional approach.

The detailed procedures of RBA are outside the scope of this book, and if interested, the readers are encouraged to research the literature.

## 5.9 Mooring software

There are several software packages available for floater and mooring dynamic analysis. The three commonly used tools in the oil and gas industry are OrcaFlex by Orcina [7], DeepC/SESAM by DNV GL [6,31,41,42], and Ariane by BV [12].

### 5.9.1 OrcaFlex by Orcina Ltd.

OrcaFlex [7] is a marine dynamic software package developed by Orcina Ltd. allowing full analysis in the time- and frequency-domain. OrcaFlex performs static and dynamic analysis of a wide range of offshore systems, typically including boundary conditions such as floating structures, buoys, etc., as well as FE modeling of line structures. The software is capable of calculating wave loads from a Morison approach. For radiation-diffraction loading, the input is needed in terms of RAOs and quadratic transfer functions. OrcaFlex solves tension, bending, and torsion using a discrete lumped mass approach with a time-stepping scheme that can be either explicit or implicit.

### 5.9.2 DeepC/SESAM by DNV GL

SESAM package is a suite of programs that includes Mimosa, Simo, Riflex, and DeepC [6,31,41,42]. It can do coupled analysis using DeepC, and frequency-domain mooring analysis using Mimosa.

DeepC is a software package consisting of two programs, Riflex and Simo. Simo is an uncoupled time-domain program for simulation of motions and station-keeping behavior of a complex system of floating structures and suspended loads. Riflex is a time-domain line dynamic simulation program developed for analysis of flexible riser systems, and it is well suited for mooring lines. In combination, DeepC is capable of analyzing the environmental impact on floating structures and mooring lines. The program is based on a nonlinear FE formulation. The lines are solved for tension, bending, and torsion and come with a linear bar element and with hybrid bar elements. The mass matrix can be lumped for efficient computations.

Mimosa is an interactive program for static and dynamic analysis of moored floating structures. It uses frequency-domain techniques to compute floater motion and dynamic mooring tension. It is capable of computing environmental loads, corresponding displacement, floater motion, and mooring tensions.

### 5.9.3    Ariane by Bureau Veritas

Ariane [12] is a mooring software developed by BV for more than 30 years. The present release, Ariane8.1, is an efficient static/time-domain multibody mooring tool based on a complex analytical solution to solve mooring lines. Floating units can be studied with up to 6 degrees of freedom per vessel and can take into account the hydrodynamic coupling between low and wave frequencies responses. Multibody hydrodynamic coupling is also part of the capabilities of Ariane8.1. Different wave drift load formulations are available to correctly evaluate waves loads in deep or shallow water or to take interaction between wave and current into account.

In addition to environmental loads, Ariane8.1 can also compute fenders, hawsers, thrusters, or user-defined loads. A dynamic positioning module is included to deal with offshore operations for example. Some of the Ariane8.1 hypotheses allow to compute a huge amount of environment cases in a relatively short time and can be used for scattering analysis. Ariane8.1 also includes a mooring line dynamic module based on Flexcom developed by Wood Group Kenny. The fatigue analysis module is also part of the software.

### 5.9.4    Other tools

Besides the above three, there are several other state-of-the-art tools:

- AQWA by ANSYS Inc. [43]
- Flexcom by Wood Group Kenny [44]
- MOSES by Bentley System [45]
- ProteusDS by Dynamic System Analysis [46]
- SeaFEM by Compass [47]
- WAMSIM by DHI Group [48]
- HARP by Texas A&M University and Offshore Dynamics Inc. [49]
- MooDy code by Chalmers University of Technology [50]

Many of these software packages include a suite of applications for the simulations of floating structures. Their capability might go beyond mooring analysis, and can be used for floater hydrodynamics and global performance analysis, riser design and analysis, installation and operation analysis, among many others.

## 5.10 Questions

1. Briefly explain what you need to model in a typical mooring analysis in order to determine the line tension and vessel offset.

2. Draw a catenary curve using a spreadsheet or any tool of your preference. Make the curve to represent a 3-in. mooring chain in 100-m water.

3. Polyester stiffness is not as straightforward as those of steel components. Why do you use one stiffness value to calculate the vessel offset and another value to calculate the line tension?

4. What is the difference between quasistatic and dynamic mooring analyses? Name one advantage of dynamic mooring analysis.

5. What is the primary assumption and limitation of frequency-domain analysis? In what kind of scenario, should a frequency-domain analysis not be used? Give an example.

6. What is the main difference between uncoupled and coupled analyses?

7. What affects the mooring system LF damping? In a time-domain analysis, the LF damping is already built-in with the mooring lines and the floater's Morison elements. How can we account for the LF damping in a frequency-domain analysis?

## References

[1] T. Sarpkaya, Experimental determination of the critical Reynolds number for pulsating Poiseuille flow, J. Basic Eng. ASME 88 (1966) 589−598.

[2] O.S. Madsen, Hydrodynamic force on circular cylinders, Appl. Ocean Res. 8 (1986) 151−155.

[3] P.A. Smith, P.K. Stansby, Impulsively started flow around a circular cylinder by the vortex method, J. Fluid Mech. 194 (1988) 45−77.

[4] K.W. Schulz, Y. Kallinderis, Three-dimensional numerical prediction of the hydrodynamic loads and motions of offshore structures, J. Offshore Mech. Arctic Eng. 122 (2000) 294−300.

[5] P.I. Johansson, A Finite Element Model for Dynamic Analysis of Mooring Cables (Ph.D. dissertation), Massachusetts Institute of Technology, 1976.

[6] MARINTEK, Riflex User Manual. Version 4.12-02, 2018.

[7] Orcina, OrcaFlex Manual. Version 9.7a. <https://www.orcina.com/>, 2013.

[8] M. Palomo, Describing Reality: Bernoulli's Challenge of the Catenary Curve and its Mathematical Description by Leibniz and Huygens, in: R. Pisano, M. Fichant, P. Bussotti, A.R.E. Oliveira (Eds.), Leibniz and the Dialogue Between Sciences, Philosophy and Engineering, 1646-2016. New Historical and Epistemological Insights, The College's Publications, London, 2017.

[9] O.M. Faltinsen, Sea Loads on Ships and Offshore Structures, Cambridge University Press, 1990.

[10] X. Xue, N.Z. Chen, Y. Wu, Y. Xiong, Y. Guo, Mooring system fatigue analysis for a semi-submersible, Ocean Eng. 156 (2018) 550−563.

[11] Y. Wu, T. Wang, O. Eide, K. Haverty, Governing factors and locations of fatigue damage on mooring lines of floating structures, Ocean Eng. 96 (1) (2015) 109−124.

[12]  BV (Bureau Veritas), VERISTAR Offshore, Aiane 8.1 Theoretical Manual. <https://www.veristar.com/>, 2018.

[13]  A. Geld, W.E. Vander Velde (Eds.), Multiple-Input Describing Functions and Nonlinear System Design, McGraw-Hill, 1968.

[14]  L.P. Kroliwkowski, T.A. Gray, An improved linearization technique for frequency domain riser analysis, in: OTC-3777-MS, Offshore Technology Conference, Houston, TX, 1980.

[15]  DNV GL, DNVGL-OS-E302, Offshore Standard Offshore Mooring Chain, 2015.

[16]  American Bureau of Shipping, Guide for Position Mooring Systems, 2018.

[17]  American Bureau of Shipping, Guidance Notes on Certification of Offshore Mooring Chain, 2014.

[18]  Massachusetts Institute of Technology Open Course, Design of Ocean Systems, MIT Course Number 2.019, Spring 2011.

[19]  API RP 2SK, Recommended Practice for Design and Analysis of Stationkeeping Systems for Floating Structures, third ed., American Petroleum Institute, 2005. Addendum 2008; Reaffirmed 2015.

[20]  ISO 19901-7, Petroleum and Natural Gas Industries − Specific Requirements for Offshore Structures − Part 7: Stationkeeping Systems for Floating Offshore Structures and Mobile Offshore Units, second ed., 2013.

[21]  DNV GL. DNVGL-OS-E301. Offshore Standard Position Mooring, 2018.

[22]  DNV GL, DNVGL-RP-C205. Recommended Practice Environmental Conditions and Environmental Loads, 2017.

[23]  American Bureau of Shipping, Guidance Notes on the Application of Fiber Rope for Offshore Mooring, American Bureau of Shipping, 2011.

[24]  C.J.M. Del Vecchio, Lightweight Materials for Deepwater Moorings (Ph.D. thesis), University of Reading, Reading, UK, 1992.

[25]  M. Francois, P. Davies, Characterization of polyester mooring lines, in: Proceedings of the ASME 27th International Conference on Offshore Mechanics and Arctic Engineering, OMAE 2008, Estoril, Portugal, 15−20 June, 2008.

[26]  M. Francois, P. Davies, Fiber rope deep water mooring: a practical model for the analysis of polyester mooring systems, in: Rio Oil and Gas Conference IBP24700, 2000.

[27]  API RP 2SM, Design, Manufacture, Installation, and Maintenance of Synthetic Fiber Ropes for Offshore Mooring, second ed., American Petroleum Institute, 2014.

[28]  C.T. Kwan, F.J. Bruen, Mooring line mooring line dynamics: comparison of time domain, frequency domain, and quasistatic analyses, in: OTC-6657-MS, Offshore Technology Conference, 1991.

[29]  BV (Bureau Veritas), 493-NR, Classification of Mooring Systems for Permanent and Mobile Offshore Units, 2015.

[30]  O.M. Aamo, T.I. Fossen, Finite element modeling of mooring lines, Math. Comput. Simul. 53 (2000) 415−422.

[31]  DNV GL, SESAM Theory Manual—DeepC Deep Water Coupled Floater Motion Analysis, Version 5.2-02, 2017.

[32]  H. Ormberg, I.J. Fylling, K. Larsen, N. Sodahl, Coupled analysis of vessel motions and mooring and riser system dynamics, in: Proc., 16th OMAE Conf., Yokohama, Japan, 1997.

[33]  H. Ormberg, K. Larsen, Coupled analysis of floater motion and mooring dynamics for a turret moored ship, Appl. Ocean Res. 20 (1998) 55−67.

[34]  M.H. Kim, Z. Ran, W. Zheng, Hull/mooring coupled dynamic analysis of a truss spar in time domain, in: Proc., Ninth ISOPE Conf., I, Brest, France, 1999.

[35]  W. Ma, M.Y. Lee, J. Zou, E.W. Huang, Deepwater nonlinear coupled analysis tool, in: Proc., OTC 2000, Paper No. 12085, Houston, TX, 2000.

[36]  Y. Luo, S. Baudic, Predicting FPSO responses using model tests and numerical analysis, in: Proc., 13th ISOPE Conf., I, Honolulu, Hawaii, USA, 2003.

[37]  A.M. Hasofer, L.C. Lind, Exact and invariant second moment code format, J. Eng. Mech. Div. ASCE 100 (1) (1974) 111−121.

[38]  S.R. Winterstein, S. Kumar, Reliability of floating structures: extreme response and load factor design, in: Proc., Offshore Technology Conference, OTC 7758, Houston, TX, May 1998.

[39]  R.G. Standing, R. Eichaker, H.D. Lawes, B. Campbell, R.B. Corr. Benefits of applying response-based design methods to deepwater FPSOs, in: Proc., Offshore Technology Conference, OTC 14232, Houston, TX, 2002.

[40]  J. Rho, J. Lee, W.-S. Sim, H.-S. Shi, C.-D. Lee, Response-based design methods for motion of turret moored FPSO, 29th International Conference on Ocean, Offshore and Arctic Engineering, Paper No. OMAE2010-20318, pp. 205−211, Shanghai, China, June 6−11, 2010.

[41]  MARINTEK, MIMOSA User's Documentation. Version 6.3-08, 2015.

[42]  SINTEF Ocean, Simo Theory Manual. Version 4.12.2, 2018.

[43]  Ansys Inc., Aqwa Theory Manual. <https://www.ansys.com>, 2015.

[44]  Wood Group Kenny, Flexcom Technical Manual, Galway, Ireland. <https://www.woodgroup.com/flexcom>, 2017.

[45]  Bentley Systems, Reference Manual for Moses, Exton, PA. <https://www.bentley.com/en/products/brands/moses>, 2015.

[46]  Dynamic System Analysis Ltd., ProteusDS 2015 Manual, Vitoria, BC, Canada. <https://dsa-ltd.ca/proteusds>, 2016.

[47]  Compass, SeaFEM Theory Manual, Barcelona, Spain. <www.compassis.com/compass/en/Productos/SeaFEM>, 2015.

[48]  DHI, Mike 21 Maritime, MIKE 21 Mooring Analysis User Guide. <https://www.mikepoweredbydhi.com>, 2017.

[49]  Offshore Dynamics Inc., HARP Manual V3. <http://www.harponline.com/>, 2003.

[50]  J. Palm, C. Eskilsson, MooDy User Manual, Chalmers University of Technology, Goteborg, Sweden, 2014.

# Chapter 6

# Fatigue analysis

## Chapter Outline

6.1 Overview    **115**
  6.1.1 Miner's rule    117
6.2 Fatigue resistance of mooring
  components    **117**
  6.2.1 $T-N$ curves for chain,
    connectors and wire ropes    118
  6.2.2 $S-N$ curves for chain
    and wire ropes    119
  6.2.3 $T-N$ curve for polyester
    ropes    120
  6.2.4 Comparison between
    $T-N$ and $S-N$ curves    121
6.3 Fatigue analysis in frequency
  domain    **122**
  6.3.1 Simple summation
    approach    123
  6.3.2 Combined spectrum
    approach    124

  6.3.3 Dual narrow band
    approach    125
6.4 Fatigue analysis in time domain    **125**
6.5 Fatigue analysis procedure    **126**
6.6 Vortex-induced motion fatigue    **128**
  6.6.1 Mechanism of
    vortex-induced motion    128
  6.6.2 Vortex-induced motion
    fatigue assessment    129
6.7 Out-of-plane bending fatigue
  for chain    **132**
  6.7.1 Mechanism of out-of-
    plane bending fatigue    132
  6.7.2 Out-of-plane bending
    fatigue assessment    134
6.8 Questions    **136**
References    **136**

## 6.1 Overview

Fatigue is a process of the cycle-by-cycle accumulation of damage in a material undergoing fluctuating stresses and strains [1,2]. A main feature of fatigue is that the load is not large enough to cause global plastic deformation or immediate failure. Instead, failure occurs after a component has experienced a certain number of load fluctuations, that is, after the accumulated damage has reached a critical level.

In general, the two-stage theory can be used to describe the process of fatigue failure [3–5]. The first stage is the fatigue crack initiation which starts from the surface of a component, where fatigue damage begins as shear cracks on crystallographic slip planes. The second stage is the crack growth which takes place in a direction normal to the applied stress, and eventually allows a fracture to occur.

Mooring System Engineering for Offshore Structures. DOI: https://doi.org/10.1016/B978-0-12-818551-3.00006-5
**115**

**FIGURE 6.1** (Left) Chain link failed due to fatigue [6]. (Right) Broken surface showing beach marks. *Courtesy: Sofec.*

Fig. 6.1 shows a broken chain link from a fatigue test (left) and the cross-section of a mooring chain link failed due to fatigue from a real incident (right). The cross-section is smooth and shows concentric rings, known as beach marks. The beach marks radiate from the origin and become coarser as the crack propagation rate increases. Each cycle of stress causes a single ripple and finally results in the whole chain link failure.

Recent offshore industry studies have found that fatigue is one of the primary reasons for offshore mooring failures, as shown in Figure 13.7 in Chapter 13, Mooring reliability. Therefore, fatigue analysis plays an essential role in the mooring design and analysis.

In general, there are two distinct approaches in fatigue analysis:

1. *$T-N$ or $S-N$ approach*—Use stress-life cumulative damage models to predict fatigue life considering the cumulative fatigue damage, where a failure occurs after a number of loading cycles $N$, at a particular tension range $T$ or stress range $S$.
2. *Fracture mechanics approach*—Use fatigue crack growth models to examine the fracture behavior of mechanical elements under dynamic loading, where failures occur if dominant cracks have grown to a critical length where the remaining strength of the component is insufficient.

The fracture mechanics approach usually is more accurate for fatigue life prediction. However, the crack growth approach is not commonly used for fatigue design in the offshore industry, mainly because of two difficulties: (1) the initial crack size is often unknown; (2) the model test data of crack versus stress are more expensive to obtain compared with $S-N$ or $T-N$ test data. This chapter focuses on the $T-N$ or $S-N$ approach with Miner's rule, which is the industry practice for fatigue analysis of mooring designs.

Numerical models with time-domain or frequency-domain dynamic analyses are used to determine tension or stress ranges. Alternatively, model test data may be used instead of dynamic analyses, provided these data are fully qualified as being suitable for fatigue analysis. The quasistatic analysis is not recommended for fatigue analysis due to its deficiency in estimating wave frequency tensions.

### 6.1.1   Miner's rule

In the stress-life cumulative damage models, Miner's rule, alternatively also called the Palmgren−Miner linear damage hypothesis, is usually applied to calculate the annual cumulative fatigue damage. Palmgren first suggested the concept of the linear damage accumulation rule in 1924 [7]. The rule was first expressed in a mathematical form by Miner in 1945 [8].

By Miner's rule, the annual cumulative fatigue damage ratio $D$ is expressed as

$$D = \sum \frac{n_i}{N_i} \tag{6.1}$$

where $n_i$ is number of cycles per year within the tension range interval $i$. $N_i$ is number of cycles to failure at normalized tension range $i$ as given by the appropriate $T-N$ or $S-N$ curve.

The design fatigue life, which is $1/D$, should be higher than the field service life multiplied by a factor of safety. For used mooring components, fatigue damage from previous operations should be taken into account.

The Miner's rule assumes that total damage caused by a number of stress cycles equals the summation of damages caused by the individual stress cycles. The primary deficiency of the Miner's rule is that it does not account for the load sequence effect, which could be significant in some situations [3,5]. However, for offshore mooring systems, the load sequence effect is usually neglected, and thus the Miner's rule is the recommended approach by industry standards and Class Rules such as those by American Petroleum Institute (API), International Standards Organization (ISO), American Bureau of Shipping (ABS) and Det Norske Veritas and Germanischer Lloyd (DNV GL) [9−12].

## 6.2   Fatigue resistance of mooring components

Strength can be easily represented by a single variable for any mooring component, whether it is for chain, wire ropes, polyester ropes or connectors. Fatigue resistance, however, is not easy to define. Resistance to fatigue can be represented by fatigue curves which are defined by a few parameters. There are two approaches to define fatigue curves, as follows:

- $T-N$ *curves*, where the tension range, $T$, is normalized by a suitable reference breaking strength (RBS), and $N$ is the permissible number of cycles.
- $S-N$ *curves*, where stress range, $S$, is defined as tension divided by the nominal cross-section area, and $N$ is the permissible number of cycles.

In either approach, the two variables are plotted on a chart as a curve to represent the fatigue resistance of a mooring component. $T-N$ curves and $S-N$ curves are determined through a regression analysis based on fatigue test data for the mooring components. The current industry practice is to use either the $T-N$ curves pre-defined in API RP-2SK codes [9] or $S-N$ curves pre-defined in DNV GL OS-E301 Class Rules [10], if no specific test data are available for the mooring component. It is noted here that both the $T-N$ curves and $S-N$ curves are developed several years ago and may not always include the latest test data, such as those for large-size chains or higher-grade material acquired in recent years [13,14]. Class societies can also approve a design curve supplied by a manufacturer for a specific type of mooring component, if the fatigue data are generated from an approved test procedure.

### 6.2.1 $T-N$ curves for chain, connectors and wire ropes

If the mooring system is designed according to API Standards, the general format of $T-N$ curves is used for calculating nominal tension fatigue lives of mooring components:

$$NR^M = K \tag{6.2}$$

This equation may be linearized by taking logarithms to give:

$$\log(N) = \log(K) - M\log(R) \tag{6.3}$$

where $N$ is number of cycles. $R$ is ratio of tension range (double amplitude) to RBS. For R3, R4, and R4S common or connecting chain links, RBS is taken as MBS (minimum breaking strength) of ORQ (oil rig quality) common chain link of the same size. For wire rope, RBS is the same as MBS. $M$ is slope of $T-N$ curve. $K$ is intercept of $T-N$ curve.

Let $L_m$ = ratio of mean load to RBS for wire rope, $M$ and $K$ values are provided below in Table 6.1.

**TABLE 6.1** $T-N$ curves $M$ and $K$ values recommended by American Petroleum Institute [9].

| Component | $M$ | $K$ |
| --- | --- | --- |
| Common studlink | 3.0 | 1000 |
| Common studless link | 3.0 | 316 |
| Baldt and Kenter connecting link | 3.0 | 178 |
| Six/multistrand rope | 4.09 | $10 \, (3.20 - 2.79 \, L_m)$ |
| Spiral strand rope | 5.05 | $10 \, (3.25 - 3.43 \, L_m)$ |

Chain fatigue is much more critical than that of wire rope. The existence of stud helps to reduce the stress concentration in stud links. Therefore, the stud link chain has better fatigue resistance than studless chain. However, in reality, the stud link chain often develops loose studs after years of use. When a stud gets loose or breaks, there will be high stress concentrations at the stud's footprint which reduce the fatigue resistance. Whether a studlink chain truly has a better fatigue life depends on these trade-off factors.

API fatigue approach recommends the safety factor of 3.0. Note that some industry practices suggest using a safety factor of 10.0 for noninspectable components and 3.0 for inspectable components [11]. Off-vessel mooring components, depending on their accessibility, are sometimes considered as noninspectable and may be required to have a safety factor of 10.0.

## 6.2.2   *S–N* curves for chain and wire ropes

If the mooring system is designed according to DNV GL Class Rules, the following equation is used to calculate the fatigue capacity of mooring line components [10]:

$$N_c S^m = a_D \tag{6.4}$$

This equation may be linearized by taking logarithms to give:

$$\log(N_c) = \log(a_D) - m\log(s) \tag{6.5}$$

where $N_c(s)$ is the number of stress cycles to fail under stress $s$. $s$ is stress range (double amplitude) in MPa defined as load divided by the nominal cross-section area. $a_D$ is the intercept parameter of the $S–N$ curve. $m$ is the slope of the $S–N$ curve.

Table 6.2 presents the $S–N$ curve parameters.

**TABLE 6.2** *S–N* curves for mooring components recommended by DNV GL [10].

|  | $a_D$ | $m$ |
|---|---|---|
| Stud chain | $1.2 \times 10^{11}$ | 3.0 |
| Studless chain | $6.0 \times 10^{10}$ | 3.0 |
| Six strand wire | $3.4 \times 10^{14}$ | 4.0 |
| Spiral strand wire | $1.7 \times 10^{17}$ | 4.8 |

The fatigue design criteria is defined by

$$1 - d_c \cdot \gamma_f \geq 0 \qquad (6.6)$$

where $d_c$ is the characteristic fatigue damage accumulated as a result of cyclic loading during the design lifetime. $\gamma_f$ is the fatigue design safety factor.

The fatigue design safety factor covers a range of uncertainties in the fatigue analysis, and is given by

$$\gamma_f = 5 \quad \text{when } d_F \leq 0.8$$

$$\gamma_f = \frac{5 + 3.0 \, (d_F - 0.8)}{0.2} \quad \text{when } d_F > 0.8$$

where $d_F$ is the adjacent fatigue damage ratio, which is the ratio between the characteristic fatigue damage $d_c$ in two adjacent lines, taken as the lesser damage divided by the greater damage. $d_F$ must be equal to or smaller than one.

The above $S-N$ curves are recommended for mooring components. DNVGL-RP-C203 also recommends different $S-N$ curves for plated structures and piles, and these curves are more generic [15].

### 6.2.3  $T-N$ curve for polyester ropes

Available data and experience indicate that there is no damage accumulation when polyester ropes are cyclically loaded with typical mooring load levels, even after millions of cycles [16]. Based on those, the industry has concluded that polyester ropes have better fatigue resistance than chain and steel wire ropes. Consequently, API recommends that the fatigue analysis for a mooring system should focus on the steel components rather than polyester ropes. When a fatigue analysis is performed for the polyester rope segments, the $M$ and $K$ values recommended in the ABS Class Rules can be used to represent the $T-N$ curve for the polyester ropes. They are listed in Table 6.3.

**TABLE 6.3** $M$ and $K$ values for polyester rope recommended by American Bureau of Shipping.

| Component | $M$ | $K$ |
| --- | --- | --- |
| Polyester | 5.2 | 25,000 |

Note that when a fatigue analysis is performed, the stiffness of the polyester ropes should be modeled in a conservative way. The storm stiffness (i.e., the highest stiffness) is used for computing the tensions in the mooring lines, because fatigue damage is mostly dominated by wave-induced motions.

Axial compression fatigue might be a failure mode for fiber ropes under compression. The industry experience has shown that axial compression fatigue is an issue for Aramid rope, but not a concern with polyester ropes, and therefore a fatigue analysis for this failure mode is not required for polyester rope [16,17].

### 6.2.4    Comparison between $T-N$ and $S-N$ curves

DNV GL provides some generic $S-N$ curves such as B1 curves for offshore steel structures as well as for out-of-plane bending fatigue or mooring connector fatigue evaluation.

Regarding mooring chain fatigue, the DNV GL $S-N$ curve and API $T-N$ curve can be compared with each other in the following way. The tension ranges, $R$, can be computed by multiplying the corresponding nominal stress range, $S$, by the nominal cross-sectional area of the components, that is, $2\pi d^2/4$ for chain, and $\pi d^2/4$ for steel wire rope, where $d$ is the component diameter. If we transfer stress into tension through chain total section area, and use ORQ common chain MBS $= [0.0211 \times D^2 \times (44 - 80 \times D) \times 10^6]$, the ratio between number of cycles to fail using $S-N$ curves and $T-N$ curves are as shown in Fig. 6.2 for the studless chain with different chain sizes.

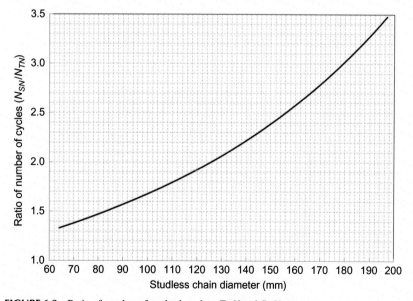

**FIGURE 6.2**    Ratio of number of cycles based on $T-N$ and $S-N$ curves.

A comparison of the API $T-N$ Curves and DNV GL $S-N$ curves is as shown in Fig. 6.3 for studless chain [18]. Overall, $S-N$ curves give higher fatigue life than $T-N$ curves, with a ratio between the two in the range of 1.3–3.5, as shown in Fig. 6.2. However, DNV GL uses a safety factor of 5.0–8.0 with the $S-N$ curves, while API use a safety factor of 3.0 with the $T-N$ curves. With the corresponding safety factors applied, the two fatigue criteria from API and DNV GL produce similar factored fatigue lives.

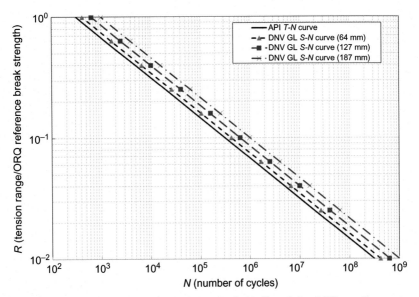

**FIGURE 6.3**   Comparisons of $T-N$ and $S-N$ curves for studless chain of different diameters.

## 6.3   Fatigue analysis in frequency domain

The most efficient way to predict the mooring system fatigue is to utilize the frequency-domain analysis to derive the tension variations for each presentative short-term sea states and then use the close-form solution to calculate the cumulative fatigue damage. In the frequency-domain approach, the wave frequency and low frequency tension variations are analyzed separately, and there are three methods to derive the overall fatigue damage, that is, the simple summation approach, combined spectrum approach, and dual narrow band approach as recommended by API, ISO, as well as class societies rules and codes [9–12].

The analysis methods of the three approaches are described in the sections below. It is noted that the time-domain analysis of mooring responses can also work with the close form solution, that is, the tension variations can also be derived from the statistics of tension time history and implemented into the close-form solution. However, in practice, when the tension time history is derived from the time-domain analysis, the cycle counting method, such as the rain-flow counting method, is usually adopted, as discussed in Section 6.4.

## 6.3.1 Simple summation approach

The simple summation method assumes that the total damage is the sum of low frequency and wave frequency fatigue damages. Wave frequency and low frequency fatigue damages for environmental state $i$, are estimated by the following equation, which is based on a Rayleigh distribution of tension peaks.

$$D_i = \frac{n_{Wi}}{K}\left(\sqrt{2}R_{W\sigma i}\right)^M \cdot \Gamma\left(\frac{(1+M)}{2}\right) + \frac{n_{Li}}{K}\left(\sqrt{2}R_{L\sigma i}\right)^M \cdot \Gamma\left(\frac{(1+M)}{2}\right) \quad (6.7)$$

where $D_i$ is annual fatigue damage from wave frequency and low frequency tensions in environmental state $i$. $n_{Wi}$ is number of wave frequency tension cycles per year for environmental state $i$. $R_{W\sigma i}$ is ratio of standard deviation of wave frequency tension range to RBS. The standard deviation of the tension range should be taken as twice the standard deviation of tension. $\Gamma$ is Gamma function. $n_{Li}$ is number of low frequency tension cycles per year for environmental state $i$. The average zero up-crossing frequency may be estimated by $1/T_n$, where $T_n$ is the natural period of the vessel computed at the vessel's mean position. $R_{L\sigma i}$ is ratio of standard deviation of low frequency tension range to RBS. The standard deviation of the tension range should be taken as twice the standard deviation of tension.

The number of tension cycles per year in each state is estimated as

$$n_i = \nu_i \cdot T_i = \nu_i \cdot P_i \cdot 3.15576 \times 10^7$$

where $\nu_i$ is the zero up-crossing frequency (hertz) of the tension spectrum in environmental state $i$. $T_i$ is the time spent in environmental state $i$ per year. $P_i$ is the probability of occurrence of environmental state $i$.

Values of the gamma function for some typical values of $m$ are given in Table 6.4.

**TABLE 6.4** Gamma functions for typical values of $m$.

| $M$ | 3.0 | 3.36 | 4.09 | 5.05 | 5.2 |
|---|---|---|---|---|---|
| $\Gamma\left[1+\frac{m}{2}\right]$ | 1.329 | 1.520 | 2.086 | 3.417 | 3.717 |
| $\Gamma\left[\frac{1+m}{2}\right]$ | 1.0 | 1.090 | 1.373 | 2.047 | 2.198 |

The simple summation method will generally give an acceptable estimate of fatigue damage if the ratio of the tension standard deviations between wave frequency and low frequency response satisfies the following condition:

$$\frac{\sigma_{WF}}{\sigma_{LF}} \geq 1.5$$

or

$$\frac{\sigma_{WF}}{\sigma_{LF}} \leq 0.05$$

where $\sigma_{WF}$ and $\sigma_{LF}$ are wave frequency and low frequency tension standard deviation, respectively.

The simple summation method may underestimate fatigue damage if both low and wave frequency tensions contribute significantly to the total fatigue damage. If both wave frequency and low frequency components are significant, the following alternatives are recommended to be used.

### 6.3.2 Combined spectrum approach

The combined spectrum approach may be used in computing the characteristic damage which is relatively simple and is always conservative. The combined spectrum method uses the standard deviation of the combined spectrum to calculate the total damage. Based on a Rayleigh distribution of tension peaks, the fatigue damage for sea state $i$ can be calculated from the following equation:

$$D_i = \frac{n_i}{K}\left(\sqrt{2}R_{\sigma i}\right)^M \cdot \Gamma\left(\frac{(1+M)}{2}\right) \tag{6.8}$$

In Eq. (6.8) the standard deviation of the combined low and wave frequency tension range, $R_{\sigma i}$, is computed from the standard deviations of the low, $R_{L\sigma i}$, and wave, $R_{W\sigma i}$, frequency tension ranges by,

$$R_{\sigma i} = \sqrt{R_{W\sigma i}^2 + R_{L\sigma i}^2} \tag{6.9}$$

The number of cycles, $n_i$, in the combined spectrum is calculated from Eq. (6.6) with the zero up-crossing frequency (hertz) of the combined spectrum, $\nu_{Ci}$, given by

$$\nu_{Ci} = \sqrt{\lambda_{Li}\nu_{Li}^2 + \lambda_{Wi}\nu_{Wi}^2} \tag{6.10}$$

where $\nu_{Wi}$ is the zero up-crossing frequency (hertz) of the wave frequency tension spectrum in environmental state $i$. $\nu_{Li}$ is the zero up-crossing frequency (hertz) of the low frequency tension spectrum in environmental state $i$ and $\lambda_{Wi}$ and $\lambda_{Li}$ are given by

$$\lambda_{Wi} = \frac{R_{Wi}^2}{R_{Wi}^2 + R_{Li}^2}, \lambda_{Li} = \frac{R_{Li}^2}{R_{Wi}^2 + R_{Li}^2} \qquad (6.11)$$

### 6.3.3 Dual narrow band approach

The combined spectrum with dual narrow-banded correction factor method uses the result of the combined spectrum method and multiplies it by a correction factor, $\rho_i$, based on the two frequency bands that are present in the tension process. The fatigue damage for environmental state $i$ is estimated from the following equation:

$$D_i = \rho_i \frac{n_i}{K} \left(\sqrt{2}R_{\sigma i}\right)^M \cdot \Gamma\left(\frac{(1+M)}{2}\right) \qquad (6.12)$$

The correction factor $\rho_i$ is given by

$$\rho_i = \frac{\nu_{ei}}{\nu_{Ci}} \left[ (\lambda_{Li})^{\frac{M}{2}+2} \cdot \left(1 - \sqrt{\frac{\lambda_{Wi}}{\lambda_{Li}}}\right) + \sqrt{\pi \lambda_{Li} \lambda_{Wi}} \cdot \frac{M\Gamma((1+M)/2)}{\Gamma((2+M)/2)} \right] + \frac{\nu_{Wi}}{\nu_{Ci}} \cdot (\lambda_{Wi})^{\frac{M}{2}} \qquad (6.13)$$

where the subscript $e$ refers to the envelope of the combined tension process, and the mean up-crossing frequency (hertz) of the envelope of the normalized tension process, $\nu_{ei}$, is given by

$$\nu_{ei} = \sqrt{\lambda_{Li}^2 \nu_{Li}^2 + \lambda_{Li} \lambda_{Wi} \nu_{Wi}^2 \delta_{Wi}^2} \qquad (6.14)$$

where $\delta_{Wi}$ is the bandwidth parameter for the wave frequency part of the normalized tension process, which may be taken as equal to 0.1.

## 6.4 Fatigue analysis in time domain

The tension time history derived from the time-domain analysis of mooring system responses can be analyzed by the rain-flow counting method to obtain the tension variation cycles. The concept of the rain-flow counting was first developed by Matsuishi and Endo [19], where the identification of cycles was likened to the path taken by rain running down a pagoda (temple) roof.

The time history of a stress (or tension) is represented as a sequence of peaks and valleys. When the time history is turned 90 degrees clockwise, the peaks and valleys look like the multiple roofs of a towered temple. The algorithm of the rain-flow counting is explained by imaging that rain drops are "spawned" between roofs. One rain drop starts at each peak as shown in Fig. 6.4 (left) and flows down. The flow path of a rain drop is counted as a half cycle. Similar half-cycles are counted for rain drops spawned at troughs, as shown in Fig. 6.4 (right). In both cases, the stress ranges for the half cycles are found as the projected distances on the stress axis. The rain-flow counting is completed by summing up all the half cycles of the same stress ranges. Note that a rain drop from a peak is allowed to flow down to the next roof below only if the next peak is shorter (e.g., peak #5 is shorter than peak #4).

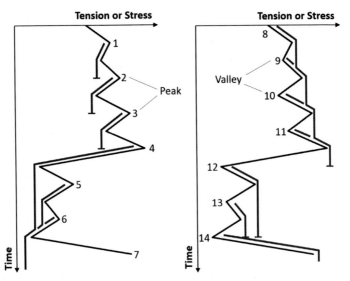

**FIGURE 6.4** Basic idea of rain-flow cycle counting, which emulates rain drops on eaves (roofs) of a pagoda.

Since the rain-flow counting approach was proposed, the method was further developed and became the commonly-used technique for cycle counting in a fatigue analysis [20].

Using the rain-flow counting method, the number of tension cycles and the expected value of the tension ranges from a time history of tension are estimated. The tension time history may be determined directly by a time-domain mooring analysis, or it may be generated from the combined low and wave frequency tension spectrum. After performing rain-flow counting, the total damage can be calculated using Miner's rule.

The rain-flow counting technique provides the most accurate estimate for fatigue damage if a sufficient number of time-domain simulations are performed that are representative of the wave scatter diagram. Note, however, it can be very time-consuming to perform a series of time-domain simulations.

## 6.5 Fatigue analysis procedure

For calculating the fatigue damage due to low frequency and wave frequency tensions, the procedure consists of the following steps.

*Step 1, Determine environmental bins*—A number of discrete bins to describe the environmental conditions are defined to represent the long term environmental conditions. Each condition consists of a direction and a sea state which is characterized by (1) significant wave height, peak period and directions, (2) current velocity profile and direction, and (3) wind velocity and direction. The probability of occurrence of each design condition is specified.

*Step 2, Run mooring analysis for each bin*—The mooring line tensions can be calculated for each defined bin as discussed in Chapter 5, Mooring analysis.

*Step 3, Determine the fatigue curve*—Determine the $T-N$ or $S-N$ curve applicable to the mooring component.

*Step 4, Compute fatigue damage for each bin*—Compute the annual damage for one environmental bin due to both low frequency and wave frequency tension using one of the methods among (1) simple summation, (2) combined spectrum, (3) dual narrow-band, or (4) rain flow counting. Repeat this step for all environmental bins and calculate the corresponding fatigue damage for each bin.

*Step 5, Sum up the fatigue damage from all bins*—Sum up the fatigue damage from all bins to get the annual fatigue damage, D. Then calculate fatigue life, L, for the mooring component.

It is important to note that the most critical fatigue location is often the top chain. The critical location, however, may change along a mooring line, depending on mooring configuration, floating structure motion characteristics, and environment. Therefore, it is recommended to assess fatigue damage along all transitional locations of a mooring line including the chain at fairlead, top of the bottom chain, and the chain at the touchdown point, as shown in Fig. 6.5 [21,22].

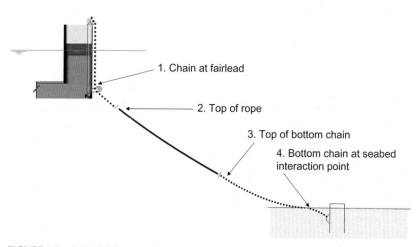

1. Chain at fairlead

2. Top of rope

3. Top of bottom chain

4. Bottom chain at seabed interaction point

**FIGURE 6.5** Critical fatigue locations along a mooring line.

## 6.6 Vortex-induced motion fatigue

The phenomenon of vortex-induced motion (VIM) on certain vessel types such as spar and semisubmersible is introduced in Chapter 3, Environmental loads and vessel motions. VIM and the associated line tension variations can contribute additional fatigue damage in mooring lines. The contribution should be added to the total fatigue damage. For floaters with slender cylindrical body shape or columns in high current areas, the contribution may be significant.

### 6.6.1 Mechanism of vortex-induced motion

Floaters can be subject to VIM if they have cylindrical hull bodies such as spars or multiple columns such as semisubmersibles and tension-leg platforms due to vortex shedding when exposed to currents (see Figure 3.9 in Chapter 3). The vortex shedding produces oscillating forces which may produce a response in any of the six rigid body modes of response, but the primary concern for the mooring design is the transverse (sway) and in-line (surge) responses. Transverse VIM occurs when the vortex shedding period is close to the transverse natural period of the floating structure, and the floating structure typically oscillates in the direction perpendicular to the current in a periodic pattern. Inline VIM is typically in the direction of the current, and it may affect the transverse VIM. The magnitude of inline VIM is typically much smaller than the transverse VIM.

Both transverse and inline motion amplitude is usually given as a nondimensional ratio of amplitude to diameter, $A/D$, where $A$ is the single amplitude of the VIM inline or transverse to the current and $D$ is the characteristic diameter of the structures. The range of current velocities over which VIM occurs can be examined using plots of $A/D$ versus the nondimensional "reduced velocity" $U_r$.

$$U_r = \frac{U_c T}{D} \tag{6.15}$$

where $T$ is the characteristic period and $U_c$ is the current velocity. An example of transverse VIM locked-in curves for a deep draft semisubmersible is shown in Fig. 6.6 for different current heading angles. In this example, VIM is negligible for all heading angles when $U_r$ is below a threshold value of 4.0. The largest VIM occurs when current directions are close to the platform diagonal direction, that is, 45 degrees, as shown in the dashed line in the figure. The range of $U_r$, where there is significant VIM, is often referred to as the "locked-in" range, in which the vortex shedding frequency may coincide with a natural frequency of the motion of the member, resulting in resonance vibrations. In this example, lock-in occurs when $U_r$ is between 6.0 and 8.0 for the current in the platform diagonal direction of 45 degrees.

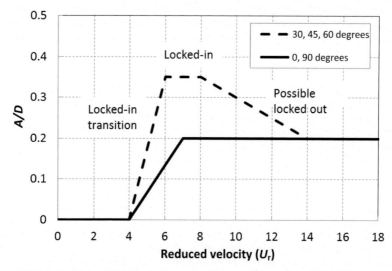

**FIGURE 6.6**  Example of VIM locked-in curves for a semisubmersible under different current headings. *VIM*, Vortex-induced motion.

Not only does VIM create vessel motions, but also it increases the drag coefficient, $C_d$, of the hull. As a result of the movement of the hull, the drag force from the current is amplified. The drag coefficient amplification factor is a function of VIM response amplitude $A/D$. The drag coefficient of the semisubmersible hull can be increased, for example, by as much as 20% in the locked-in phase, compared with the drag coefficient without VIM.

Both $A/D$ and $C_d$ are a function of a large number of parameters such as reduced velocity, platform shape, strake configuration in case of spars, current characteristics, and hull appurtenances, etc. The common industry practice is to perform model tests to obtain $A/D$ and $C_d$ curves. However, model tests can only model certain parameters while approximating others. Therefore, cautions should be taken in the interpretation and use of model test data [9,23–27].

## 6.6.2  Vortex-induced motion fatigue assessment

Mooring tensions due to VIM are cyclic by nature and contribute to fatigue damage of the mooring system. API RP 2SK [9] describes an analysis procedure for long-term VIM fatigue damage evaluation. The procedure focuses on fatigue of the chain at the fairlead location. However, it is recommended to assess fatigue damage along all transitional locations of a mooring line following the similar procedure. The procedure is summarized as the following steps, based on API RP 2SK [7,9] and recent industry practices [18,28].

*Step 1, Determine current events*—The long-term current events can be represented by a number of discrete current bins. Each current bin consists of a reference direction and a reference current velocity with the associated wave and wind conditions. The probability of occurrence of each current bin is specified.

*Step 2, Select a current bin and get its duration*—Select a current bin and calculate the duration $t_i$ for the current bin in a year based on the probability of occurrence for the current velocity and direction.

*Step 3, Determine A/D and Cd through iteration*—Determine the natural period $T_i$ of the moored floater under the current bin without VIM, based on an estimated drag coefficient $C_d$.

Step 3.1: Specify extreme in-line and transverse A/D values for the current bin based on available model test or field measurement data. The mean A/D for fatigue analysis can be evaluated by multiplying the extreme A/D by a coefficient g, which should be determined by the available model test or field measurement data.

Step 3.2: Determine in-line and transverse VIM amplitude coefficient $C_v$, which is a function of reduced velocity, and is equal to 1.0 at peak VIM under lock-in condition.

Step 3.3: Calculate the reduced velocity for the current bin and further modify the mean in-line and transverse A/D by $C_v$.

Step 3.4: Determine the drag coefficient $C_d$ for the current bin based on the modified mean transverse A/D. If this $C_d$ value is significantly different from the estimated $C_d$, iteration may be required.

*Step 4, Run mooring analysis to obtain tension range*—Perform mooring analysis based on the modified (updated) mean in-line and transverse A/D (Step 3.3), and $C_d$ (Step 3.4), using the analysis procedure for strength design. Determine average tension ranges $R_i$, and corresponding average response period $T_i$ from the time trace of line tensions for a few VIM cycles.

*Step 5, Calculate fatigue damage for the tension range using T–N curve*—Determine the number of cycles to failure, $N_i$, corresponding to tension range, $R_i$, for the mooring component of interest using an appropriate T–N equation/curve. Calculate annual fatigue damage for the i-th current bin:

$$D_i = \frac{(t_i/T_i)}{N_i} \tag{6.16}$$

*Step 6, Repeat for other bins, and sum up the fatigue damages*—Repeat Steps 2–5 for other bins. Sum up the damages from all bins to get the fatigue damage due to VIM.

Fig. 6.7 shows an example of fatigue damage $D_i$ of different current speeds. VIM fatigue damage is to be combined with the fatigue damage from wind and waves to obtain total fatigue damage of the mooring system. The predicted fatigue life is $1/D_i$ (years), which should be greater than the service life times a factor of safety.

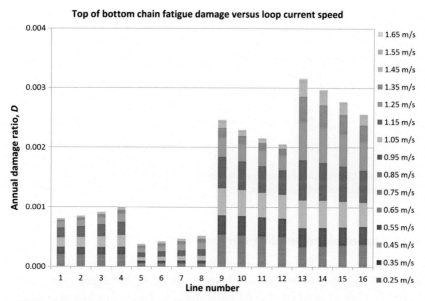

**FIGURE 6.7** Example of computed VIM fatigue damages contributed by different current events for 16 mooring lines. *VIM*, Vortex-induced motion.

For some mooring systems, considerable fatigue damage may be caused by a single extreme VIM fatigue event, which should also be addressed. Note that the current for the worst-case VIM fatigue event may not coincide with the 100-year return period current event, but could occur under lower return period current events. For fatigue analysis of single VIM events, the current criteria should specify the current velocity, profile, direction, and duration (build-up and decay) for current events spanning a range of return periods [9].

It is highly recommended to use a dynamic analysis approach for mooring line tension. Even though a VIM motion has a long natural period in the range of approximately 100−200 seconds, the quasistatic analysis may still underpredict the tension ranges because of the neglection of line dynamics. For instance, Fig. 6.8 shows an example of one-line tension ranges at the fairlead location using dynamic and quasistatic analysis approaches. In this example, the quasistatic analysis approach underestimates tension range by 13% compared with the dynamic analysis approach. The resultant fatigue damage is underestimated by around 44%.

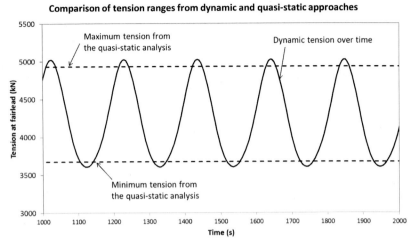

**FIGURE 6.8** Comparison of tension ranges from dynamic and quasistatic approaches for one mooring line under one VIM current bin.

## 6.7 Out-of-plane bending fatigue for chain

Since the first out-of-plane bending (OPB) fatigue failure was discovered in the mooring legs of the Girassol buoy in 2002, OPB fatigue of mooring chains has been identified as a potential mooring failure mechanism and has drawn some attention [29−32]. OPB fatigue has been confirmed to be the main root cause of at least three mooring failures [30]. This section briefly discusses the mechanism of OPB fatigue and summarizes current analysis methodology.

### 6.7.1 Mechanism of out-of-plane bending fatigue

For chain with smooth interlink contact surfaces, it was the general under-standing of mooring practitioners that two adjacent links can rotate around each other due to interlink rolling and sliding. It was later found that links can lock into each other especially under high pretension in deep waters. Fig. 6.9 shows the interlink locking that happened to chain links inside the hawse pipe of Girassol buoy [29,30,32].

During the chain manufacturing process, chain links are typically proof loaded to 70%−80% of their MBS. The proof-loading leads to plastic defor-mation of the chain links, especially in the grip area between the links. This change of geometry due to proof loading will introduce interlink rotational stiffness and in simple words will cause the links to lock into each other. As a result, instead of two phases with rolling and sliding, the relationship

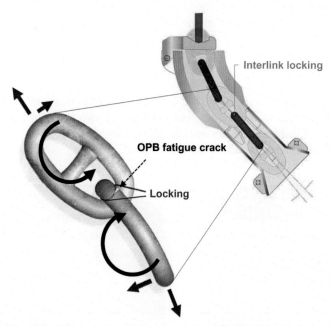

**FIGURE 6.9**   Chain OPB mechanism inside a hawse pipe [29,30]. *OPB*, Out-of-plane bending.

between bending moment and interlink angle can be divided into three phases [29]: locking, stick-sliding, and sliding.

1. Locking: In this phase, the chain links are locked with each other without any relative motion in the contact area. The chain links behave as a single rigid beam element. The bending moment increases linearly as the interlink angle increases. The slope of the bending moment versus interlink angle in the locking phase is known as the interlink stiffness.
2. Stick-sliding: This phase can be considered as a transition phase between locking and sliding. The relationship between bending moment and interlink angle becomes more nonlinear in this phase compared with that in the locking phase.
3. Sliding: The relative motion of the adjacent links is characterized by sliding at the contact area. In this phase, the bending moment remains constant with the increase of the interlink angle.

During the locking and stick-sliding phases, the magnitude of the bending stress is significantly higher than the bending stress that develops during rolling phase in nonproof-loaded chain links. When the vessel motion causes OPB in the mooring chain, the bending in the chain link amplifies the stress variation and eventually can lead to OPB fatigue failure. In theory, OPB fatigue can occur at any location where adjacent chain links undergo relative angular movement. However, the most significant relative angles are typically found in the few links at the connection between the chain and the floater.

Note that the crack initiation points of OPB fatigue and tension−tension (TT) fatigue are at distinctly different locations on a chain link. Fig. 6.10 compares the crack initiation points, shown as hot spots, between OPB and TT fatigue. In the case of pure OPB loading, the hotspots are located in the bend region, close to the contact area between the two links. In the case of pure tensile loading, there are two distinct locations of hotspots: the inner bends (also known as intrados) and the crowns.

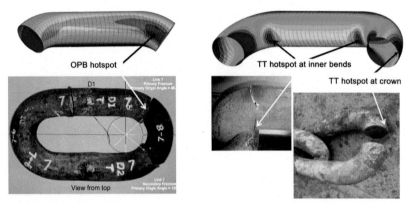

**FIGURE 6.10** Comparison of hot spot locations between OPB loading and TT loading [30]. *OPB*, Out-of-plane bending; *TT*, tension−tension.

## 6.7.2 Out-of-plane bending fatigue assessment

Compared with the traditional TT fatigue of mooring chain which has been studied for many years, OPB fatigue is relatively novel. Its analysis involves multiple disciplines including mooring, structural, and mechanical engineering. The out-of-plane bending of a mooring chain involves a complex mechanism so that it is difficult to determine whether the OPB fatigue must be assessed for a particular design of the fairlead or hawse pipe. The need to conduct OPB fatigue analysis relies on practitioners' experience and engineering judgment.

Some guidance has been published that recommends and summarizes OPB assessment methodologies [30,32]. However, they are not considered mature recommended industry practice at this point, and further investigation and verification may be needed. Overall, the OPB fatigue analysis procedure is somewhat similar to the fatigue analysis of a structural component. The OPB fatigue analysis is performed in the following main steps:

*Step 1, Develop fatigue sea states*—The process of developing OPB fatigue sea states is similar to what is typically done for TT fatigue analysis. However, OPB fatigue damage is typically more evenly distributed among different environmental combinations as compared to TT fatigue damage [29]. Therefore, a sea state matrix with larger number of environmental combinations and higher bin resolution is usually required for the OPB fatigue evaluation purpose.

*Step 2, Develop interlink stiffness and stress concentration factors (SCF)*—The chain interlink bending stiffness describes the relationship between the interlink angle and the nominal bending moment generated between two adjacent links. The finite element analysis (FEA) can be used to estimate the bending stiffness and SCF at OPB hotspots. Full-scale chain testing can also be used to estimate interlink stiffness. In the early design phase without conducting FEA analysis and chain testing, the interlink stiffness model and SCF recommended by Bureau Veritas (BV) can be considered [31]. In the BV stiffness model, the bending moment is estimated as an empirical function of interlink angle, mooring line tension, and chain diameter. The SCF is based on hotspot locations. At OPB hotspot location the SCF is recommended as a function of chain pretension level with a minimum value of 1.15.

*Step 3, Perform global response analysis and local modeling*—The objective of global response analysis and local chain link and connection modeling is to estimate the time-series of tension and bending moment components of the chain links in a specific fatigue sea state. The analysis comprises two parts. First, perform a global vessel-line response analysis which estimates the time-series of line tension and relative line-connection total angles. The analysis is typically done in the time domain due to the numerical modeling requirements and the complexity of the analysis process. Second, model the local chain connection to transfer the total angles to the local interlink angles and moments between the links. The modeling is typically done using a simplified FEA model of the chain segment and top connection.

*Step 4, Calculate total stress and count cycles*—After the global response analysis, stress calculation, and cycle counting, the time-series of tension and primary and secondary moment components are used to calculate the nominal tensile, OPB, and in-plane bending (IPB) stress components in the affected links using the moments of area of the chain link. The rainfall counting approach is applied on the total stress time-series to develop the stress range histogram of each sea state. The long-term stress range histogram is developed from the stress range histogram of each sea state and the corresponding probability of occurrence of the sea state. The total fatigue damage is calculated based on the $S-N$ curve approach with Miner's rule. The DNV GL B1 $S-N$ curve [15] is a popular choice for OPB fatigue analysis.

## 6.8 Questions

1. Briefly explain the difference between the $T-N$ curves and $S-N$ curves. The two sets of curves come with different fatigue safety factors. Explain why the difference in safety factors may not be an issue.
2. What are the approaches for predicting fatigue damages? Which is always conservative? Which gives the most accurate fatigue estimation?
3. What technique for cycle counting is widely used in time-domain fatigue analysis?
4. Describe the mechanism of VIM. How does VIM affect mooring fatigue?
5. Describe the mechanism of chain OPB. What location on a mooring line is vulnerable to OPB fatigue damage?

## References

[1] A. Almar-Naess, Fatigue Handbook: Offshore Steel Structures, Tapir Academic Press, Trondheim, Norway, 1985.

[2] A. Naess, T. Moan, Stochastic Dynamics of Marine Structures, Cambridge University Press, 2012. ISBN: 9781139021364.

[3] W. Cui, A state-of-the-art review on fatigue life prediction methods for metal structures, J. Mar. Sci. Technol. 7 (2002) 43–56.

[4] P.J.E. Forsyth, The Physical Basis of Metal Fatigue, American Elsevier, New York, 1969.

[5] J. Schijve, Fatigue of Structures and Materials, Springer, 2009.

[6] K. Ma, Ø. Gabrielsen, Z. Li, D. Baker, A. Yao, P. Vargas, et al., Fatigue tests on corroded mooring chains retrieved from various fields, in: OMAE2019-95618, June (2019) 9–14.

[7] A. Palmgren, Die lebensdauer von kugellagern, 68, Verfahrentechinik, Berlin, 1924, pp. 339–341.

[8] M.A. Miner, Cumulative damage in fatigue, J. Appl. Mech 12 (1945) A–159.

[9] API RP 2SK, Recommended Practice for Design and Analysis of Stationkeeping Systems for Floating Structures, third ed, American Petroleum Institute, October 2005. Addendum 2008; Reaffirmed 2015.

[10] DNV GL, DNVGL-OS-E301. Offshore Standard Position Mooring, 2018.

[11] American Bureau of Shipping, Guide for Position Mooring Systems, 2018.

[12] ISO 19901, Petroleum and Natural Gas Industries—Specific Requirements for Offshore Structures—Part 7: Stationkeeping Systems for Floating Offshore Structures and Mobile Offshore Units, 2013.

[13] J. Fernandez, W. Storesund, J. Navas, Fatigue performance of grade R4 and R5 mooring chains in seawater, in: Proceedings of the 33rd International Conference on Ocean, Offshore and Arctic Engineering, OMAE 2014-23491, San Francisco, June 2014.

[14] Noble Denton & Associates, Inc, Corrosion Fatigue Testing of 76mm Grade R3 & R4 Studless Mooring Chain, H5787/NDAI/MJW, Rev 0, 2002.

[15] DNV GL, DNVGL-RP-C203, Fatigue Design of Offshore Steel Structures, 2016.

[16] API RP 2SM, Design, Manufacture, Installation, and Maintenance of Synthetic Fiber Ropes for Offshore Mooring, second ed, American Petroleum Institute, 2014.

[17] American Bureau of Shipping, Guide Notes on the Application of Fiber Rope for Offshore Mooring, 2014.

[18]  Y. Wu, T. Wang, R. Eide.Governing locations of offshore mooring fatigue design, in: Proceeding of the 20th SNAME Offshore Symposium, SNAME_OS15_06, Houston, TX, 2015.

[19]  M. Matsuishi, T. Endo, Fatigue of Metals Subjected to Varying Stress, Japan Society of Mechanical Engineers, Fukuoka, 1968.

[20]  G. Marsh, C. Wignall, P.R. Thies, N. Barltrop, A. Incecik, V. Venugopal, et al., Review and application of Rainflow residue processing techniques for accurate fatigue damage estimation, Int. J. Fatigue 82 (3) (2016) 757−765.

[21]  X. Xue, N.Z. Chen, Y. Wu, Y. Xiong, Y. Guo, Mooring system fatigue analysis for a semi-submersible, Ocean Eng. 156 (2018) 550−563.

[22]  Y. Wu, T. Wang, O. Eide, K. Haverty, Governing factors and locations of fatigue damage on mooring lines of floating structures, Ocean Eng. 96 (1) (2015) 109−124.

[23]  M.B. Irani, L.D. Finn, Model testing for vortex induced motions of spar platforms, in: Proceeding of the 23rd International Conference on Offshore Mechanics and Arctic Engineering OMAE 2004-51315, 2004.

[24]  R.T. van Dijk, A. Voogt, P. Fourchy, M. Saadat, The effect of mooring system and sheared currents on vortex induced motions of truss spars, in: Proceedings of the 22nd International Conference on Offshore Mechanics and Arctic Engineering, OMAE'03, Cancun, Mexico, 2003.

[25]  K. Huang, X. Chen, C.T. Kwan, The impact of vortex-induced motions on mooring system design for spar-based installations, in: Offshore Technology Conference, 15245, Houston, TX, 2003.

[26]  T. Kokkinis, R.E. Sandström, H.T. Jones, H.M. Thompson, W.L. Greiner, Development of a stepped line tensioning solution for mitigating VIM effects in loop/eddy currents for the genesis spar, in: Proceeding of the 23rd International Conference on Offshore Mechanics and Arctic Engineering, OMAE2004-51546, 2004.

[27]  G. Wu, W. Ma, M. Kramer, J. Kim, H. Jang, J. O'Sullivan, Vortex induced motions of a column stabilized floater, Part II: CFD benchmark and prediction, in: Deep Offshore Technology International Conference, Aberdeen, Scotland, October 2014, 2014.

[28]  Y.C. Park, A. Antony, H. Moideen, A. Jamnongpipatkul, Gulfstar—VIM and mooring chain fatigue, in: OTC-26051-MS, Offshore Technology Conference, Houston, TX, 2015.

[29]  P. Jean, K. Goessens, D. L'Hostis, Failure of chains by bending on deepwater mooring systems, in: OTC 17238, Offshore Technology Conference, Houston, TX, 2005.

[30]  A. Izadparast, C. Heyl, K. Ma, P. Vargas, J. Zou, Guidance for assessing out-of-plane bending fatigue on chain used in permanent mooring systems, in: Proceedings of the 23rd Offshore Symposium, Society of Naval Architects and Marine Engineers (SNAME), Houston, February 2018.

[31]  Bureau Veritas, Fatigue of Top Chain of Mooring Lines Due to In-Plane and Out-of-Plane Bending, Guidance Note NI 604 DT R00 E, 2014.

[32]  L. Rampi, F. Dewi, P. Vargas, Chain out of plane bending (OPB) joint industry project (JIP) summary and main results, in: Offshore Technology Conference, 25779-MS, Houston, TX, 2015.

Chapter 7

# Model tests

## Chapter Outline

**7.1 Types of model tests**      **140**
   7.1.1 Ocean basin model test      140
   7.1.2 Wind tunnel test      141
   7.1.3 Towing tank test      141
   7.1.4 Ice tank test      142
**7.2 Principle of model test**      **143**
   7.2.1 Scale factor      144
**7.3 Capability of model basin facilities**      **145**
   7.3.1 Wind generation      145
   7.3.2 Wavemaker      145
   7.3.3 Current generation      146
**7.4 Limitations of model test**      **147**
**7.5 Mooring system truncation**      **148**
   7.5.1 Mooring truncation      148
   7.5.2 Truncation design      149

   7.5.3 Limitations due to
      truncation      149
   7.5.4 Other truncation methods      150
**7.6 Hybrid test method**      **150**
   7.6.1 Hybrid method      150
   7.6.2 Basic principle      150
   7.6.3 Numerical tools      151
**7.7 Model test execution**      **152**
   7.7.1 Model preparation      152
   7.7.2 Environment calibration      152
   7.7.3 Data collection and
      processing      152
**7.8 Questions**      **153**
**References**      **153**

Model tests are often used for the design verification of floating structures and their mooring systems. In addition to advanced analysis of the mooring system responses, the model test is a useful design and verification tool which is used for the following purposes:

- Verify the overall behavior of the floating system.
- Determine the design values of loads and responses.
- Validate and calibrate numerical analysis tools.
- Discover any unexpected effect that was not yet taken into account in the design.

In addition to the above purposes, model tests can also be utilized to determine responses due to nonlinear phenomena that cannot be reliably predicted by analytical methods, such as wave runup, underdeck wave slamming, fish-tailing, and vortex-induced motion.

Mooring System Engineering for Offshore Structures. DOI: https://doi.org/10.1016/B978-0-12-818551-3.00007-7

## 7.1   Types of model tests

Model tests place physical models of a reduced scale under simulated environments in man-made test facilities. By measuring the motions and loads of the reduce scale model, the responses of the prototype floating system can be predicted. There are many types of model tests serving different purposes. In the present book, only those tests that are directly related to mooring system design are described.

### 7.1.1   Ocean basin model test

The test facility that is mostly relevant with the design of mooring systems is the ocean basin, which is also called offshore basin or model basin. The ocean basin is widely used to measure the offset, motion responses, and hydrodynamic loads of a floating system. The ocean environment including the waves, wind, and current can be simulated, and the responses of the physical model (either free-floating or moored to the basin bottom) are measured. A model basin typically has sufficient size and depth so that tests can be conducted at a reasonable scale. The facilities are capable of generating regular and random waves over a wide frequency range, and steady and turbulent wind and defined current profile (Fig. 7.1).

**FIGURE 7.1**   Model test in an ocean basin. *Courtesy: SINTEF/NTNU.*

The test program at a model basin typically is configured to proceed through a sequence of tests, such as the following:

- *Static offset tests*—To derive and verify the mooring stiffness characteristic.
- *Decay tests*—To verify the loading conditions and derive the floater natural periods and damping for free-floating and moored conditions.
- *Sea-keeping tests*—To derive the motion transfer functions.
- *Station-keeping tests*—To obtain the floater offsets and mooring/riser tensions.

## 7.1.2    Wind tunnel test

A wind tunnel is a testing facility frequently used to study the wind (and current) loads on offshore structures. A wind tunnel consists of a tubular passage with the object under test mounted in the middle. Air is made to move past the object by a powerful fan system. The test object is tilted and rotated to measure the aerodynamic forces, pressure distribution, or other aerodynamic-related characteristics (Fig. 7.2).

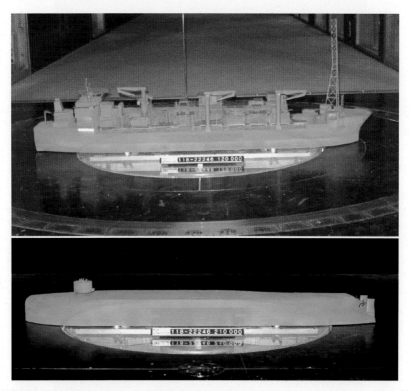

**FIGURE 7.2**    Wind tunnel test. Top: hull and topsides for wind force test. Bottom: Hull is upended for current force test. *Courtesy: NOV/APL.*

## 7.1.3    Towing tank test

A towing tank is used to carry out hydrodynamic tests of ships or platform models (e.g., semisubmersible) to obtain hydrodynamic properties to be used in the new design or modification of floating structures. The towing tank is of long and narrow shape and has a rail track for towing the model at various speeds. It is often used to derive the drag force coefficients of the floater and can also be used for vortex-induced vibration testing (Fig. 7.3).

**FIGURE 7.3**   Towing test in progress. *Courtesy: SINTEF.*

## 7.1.4   Ice tank test

An ice tank is used to test ice loads on ships and offshore platforms operating in icing conditions. The tank is like a gigantic refrigerator to simulate ice conditions. When a vessel or platform is towed in simulated ice thickness, coverage, and hardness, ice forces on offshore structures can be determined. Ice layers are frozen with a special procedure to scale down the ice crystals to model scale (Fig. 7.4).

**FIGURE 7.4**   Ice tank test. *Courtesy: SOFEC.*

## 7.2  Principle of model test

Experimental testing of scale models is a standard tool in offshore engineering. The full-scale platform (i.e., prototype) can have a dimension of over 100 m, while the physical model for testing is only a fraction of the size. The beauty of model testing is the application of scale laws of "similarity" to predict the responses of a prototype (full-scale structure) through testing the reduced scale model. There are the following three types of similarity:

- *Geometrical Similarity* means that the model and the full-scale structure must have the same shape, and all linear dimensions must have the same scale ratio.
- *Kinematic Similarity* (*Similarity of velocity*) means that the flow and model will have geometrically similar motions.
- *Dynamic Similarity* means similarity of forces, and ratios between different forces in full scale must be the same in model scale.

Based on the above similarities, there are a number of well-known force ratios. Two of these dimensionless numbers are listed here as examples:

- Froude number $F = (\text{inertial force/gravity force})^{1/2} = U/(gL)^{1/2}$
- Reynolds number $R = \text{inertial force/viscous force} = UL/v$

For phenomena where gravity and inertial forces are dominant and the effects of remaining forces, such as kinematic viscosity, $v$, are small, the model test would match the Froude number. Based on this principle, the model and its prototype can be related by the factors as listed in Table 7.1. The scale factor, $\lambda$, is defined as the linear dimension (length) ratio of the prototype over the model, $\rho_m$ is fluid density in model scale, and $\rho_f$ is fluid density in full scale.

**TABLE 7.1 Froude scale factors.**

| Physical parameter | Unit | Multiplication factor |
| --- | --- | --- |
| Length | m | $\lambda$ |
| Mass | kg | $\lambda^3 \rho_f/\rho_m$ |
| Force | N | $\lambda^3 \rho_f/\rho_m$ |
| Moment | N m | $\lambda^4 \rho_f/\rho_m$ |
| Acceleration | m/s$^2$ | $a_f = a_m$ |
| Time | s | $\lambda^{1/2}$ |
| Pressure | N/m$^2$ | $\lambda \rho_f/\rho_m$ |

Note that it is impossible to have all identical force ratios between model and its prototype, and during model tests, one can only be selective such that the most relevant force ratios are identical between the testing model and its prototype and the remaining result in scale effects. When modeling according to Froude scale, viscous forces (which are related to Reynolds number) are not exactly matched.

In summary, model tests use reduced scale models based on the law of similarity, and the responses of motion and load measured during tests can be converted to prototype scale based on the scale factors.

## 7.2.1 Scale factor

With respect to the choice of the scale factor of model testing, there are some considerations. The larger the model is (i.e., the smaller the scale factor is), the more reliably one can model the free surface interactions with the floater. The smaller the model is, the more difficult it is to achieve good equivalence between the model and its prototype. With very small models (e.g., $\lambda > 100$), some practical problems begin to surface for the conduct of model tests. The wave heights become very small and basin noise generated by unsteady wavemakers, air currents, etc. begins to become significant. Also, models become very light in weight, and it becomes more difficult to match mass properties (Fig. 7.5).

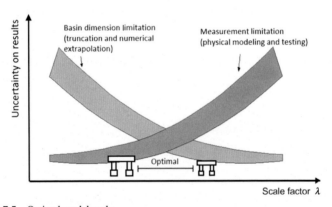

**FIGURE 7.5** Optimal model scale.

Typically, the scales of truncated systems are in the range from 1:40 to 1:90. Such scales have the following advantages:

- Inviscid wave effects at the free surface are modeled at an acceptable scale.
- Acceptable or manageable scale effects, for instance of the mooring lines and risers.

- Accurate measurements are possible due to the accurate modeling of dimensions and weight distributions.
- Good quality wind, wave, and current generation is possible.

## 7.3    Capability of model basin facilities

### 7.3.1    Wind generation

The wind environment at the model test basin is usually generated by a group of fans. Based on the target wind load, the wind speed is calibrated to achieve the right wind load. Ideally the wind generation fans can move freely (rotation and translation movement) around the test model to achieve the optimum wind environment. Note that it is always challenging to create a field of a uniform wind speed across the area surrounding the model even with a large number of fans (Fig. 7.6).

**FIGURE 7.6**    Wind generation. *Courtesy: MARIN.*

### 7.3.2    Wavemaker

Wave generation capability is one of the key features of a model basin. A state-of-the-art basin should be fitted with a wavemaker that is capable of generating regular waves, random waves including white noise waves, wave groups, and multidirectional wave. Square or rectangular wave basins are usually equipped with banks of single flip or multiflip wavemakers on one or two sides of the basin. Beaches are used on the opposite sides of the wavemakers to minimize undesired reflected waves. In most cases, beaches consist of sloping porous surfaces which absorb the wave energy through a combination of viscous dissipation and breaking. However, partial reflection from beaches is inevitable, so in some cases it is necessary to limit the duration of an experiment so that the reflected waves do not affect the test area around the model.

For generating quality waves, the basin should have adequate size to minimize the hydrodynamic effects associated with the finite dimensions of the basin. Wall effects which are not related directly to waves would be negligible if the basin is sufficiently large relative to the model length scale. However, the effects associated with wave reflection from the walls will persist regardless of the size of the basin. Thus, it is essential to both generate and absorb waves in a controlled manner, to simulate the open-water conditions (Fig. 7.7).

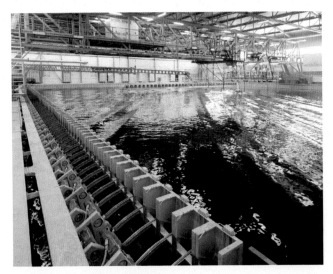

**FIGURE 7.7** Wavemaker. *Courtesy: MARIN.*

### 7.3.3 Current generation

To generate a uniform current profile in a model test basin is challenging. In order to generate a current, a large number of pumps are used at different pumping rates across the basin depth to generate the flow. To have steady flow across the basin, the flow has to be circulated. In some basins, the flow is circulated beneath the bottom floor of the basin. The best circulation with minimum interference with waves is to circulate the flow outside the basin. However, even with the most advanced current generation facilities, it is not possible to achieve a perfect uniform current profile. The presence of current turbulence in the order of 5%−10% is inevitable. Some model basins have no current generation capability, and sometimes a group of local fans can be positioned inside the basin to generate local current. It should be noted that it is even more difficult to control the quality of locally generated current.

## 7.4    Limitations of model test

Even though model testing is a powerful tool for predicting the dynamic responses of the floater, it has a few limitations, that is, (1) theoretical limitation; (2) basin dimension limitation; (3) facility capability limitation; and (4) test case limitation.

1. *Theoretical Limitation*—It is not possible to simultaneously satisfy the scaling of both Froude and Reynolds numbers. As a result, the quality of similarity cannot be achieved between the various forces. In other words, the ratios between the various forces will be different at model scale and prototype scale. Therefore, choices must be made based on the relative importance of the forces to the problem at hand. Since wave forces are generally the most important forces in model testing of floaters and are related to gravity forces, the *Froude number* is the most important dimensionless number and is used as the basis for scaling model tests of moored floaters. In applying the Froude scaling law, "scale effects" are introduced because of the noncompliance with Reynolds number similitude. The most important scale effect in model tests of moored floaters is that due to viscous drag effects. The correction factors needed are unknown, and such compensation suggests a larger test accuracy than realistic. The well-known dependency of the drag coefficient (Cd) value on the Reynolds number in a steady flow is not valid for the complex flow processes that occur during a test. If the coupled mooring analysis of the system indicates sensitivity to this type of scale effect, it is recommended that separate "component" tests are performed at appropriate Reynolds numbers.

2. *Basin Dimension Limitation*—Ideally, the whole floater and its mooring and riser model are well fitted in the test basin. However, test basins have limited dimensions both in size and depth. For a deep-water floating system designed for a water depth of 3000 m, a very-deep basin with a 10-m depth can only model the complete system using a scale of about 1:300, which is far too small. Due to the limitation of the physical size of the basin, the mooring legs and risers have to be truncated. It then brings the problem of designing a truncated model mooring system to be representative of the untruncated system.

3. *Facility Capability Limitation*—The quality of wave-making, current, and wind generation are limited by the facilities' capabilities. For example, the basin's wave-making capability is typically limited to a certain height, period, and steepness. Also, the current profile is supposed to be uniform in theory, but the basin-generated current speed will inevitably have turbulence. Additionally, the wind is generated by a group of fans, but the wind speed is not likely to be completely uniform across the testing field.

**4.** *Test Case Limitation*—Model testing is time-consuming. From the calibration of environmental parameters to the trial test and to the final test, it can take many days. The basin time is valuable, and as a result only limited test cases can be conducted during a model test campaign. It is often required that extensive engineering analyses are conducted beforehand to shortlist critical cases, and only these cases are tested in the basin.

## 7.5 Mooring system truncation

### 7.5.1 Mooring truncation

To conduct a model test, it is required to create a representation of the entire floating system including the mooring lines and risers. In deep water, the floating system has very long mooring lines in the water column. A model basin may not be able to accommodate the complete model of the floating system with a reasonable scale simply due to the limitation of the basin's physical dimension. Therefore, the common practice is to employ a truncated mooring system with as much equivalence to the prototype mooring system as possible, as illustrated in Fig. 7.8.

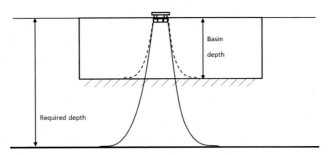

**FIGURE 7.8** Illustration of equivalent (truncated) mooring lines.

The terms "truncated mooring system" and "equivalent mooring system" have been used by different groups to mean essentially the same thing. Truncation means any reduction in the size of the mooring and riser system due to basin depth or spatial limitations. The portion of the mooring or riser system below a truncation depth or width may be ignored or replaced with a passive or active mimic device. Equivalence means the truncated model should have equivalent load—displacement characteristics as the full-depth model which is not feasible due to the basin's spatial limitation; and the models mimic the behavior of the prototype floating system.

## 7.5.2    Truncation design

When it is necessary to employ truncated (equivalent) systems, the objective of the truncation design is to assure that the behavior of a particular floater type at the free surface is as close as possible to that of the full-depth system. This allows the direct evaluation of design issues from the model test results.

An underlying problem in truncation is that one would like to maintain the same forces, motions and offset of the full-depth system while dealing with the constraint of a reduced water depth and a reduced anchor radius. This implies that the geometrical stiffnesses in the system stiffnesses will be too large due to the reduced geometry. A commonly adopted method to overcome this problem is to introduce elastic compliancy by adding coil springs.

To conduct a truncation design, the first step is to create a truncated (equivalent) mooring system that has the static load—displacement characteristics matched as closely as possible to that of the full-depth system. An initial truncation design can be based on the concept of geometric scaling of the line shapes and keeping the catenary effects of the top and bottom chain intact at the same time. An alternative method is to select the anchor radius to be as large as possible in the basin, in order to compensate for the lack of mooring line length. Regardless of which method is used, the correct modeling of mooring horizontal stiffness is most important.

As a next step, the static load—displacement curve of the truncated system is tuned and optimized. For a catenary moored system, the static load—displacement curve has three regions: (1) where the catenary effect dominates the system stiffness; (2) where the elasticity in the mooring line dominates the system stiffness; and (3) a transition region between the two regions. The truncation implies that the region of geometric effect is reduced, and the region of elastic deformation is increased. To reflect the nonlinearity of the catenary effect by elastic deformation, it may be necessary to use nonlinear (blocked) springs or increased catenary line weights. As the final step, the dynamic response of the truncated system is checked to ensure a close match to the full-depth system.

## 7.5.3    Limitations due to truncation

It is important to note that any truncated (equivalent) system has its limitations. It is not possible to precisely reproduce all quasistatic and dynamic characteristics (in all six degrees of freedom) of a full-depth system with a truncated equivalent system. When it is possible to design a truncated mooring system that follows the behavior of the full-depth system, this is then limited to a certain displacement range (i.e., vessel offset range around a zero, mean, or expected extreme positions).

### 7.5.4 Other truncation methods

The mooring and riser truncation techniques presented in the previous sections are those commonly practiced by the model basins and well accepted by the offshore industry. Depending on the purpose and acceptance criteria of testing, there are alternative truncation techniques.

One example is to truncate the prototype moorings and risers at the basin floor and develop an active or passive system that mimics the response of the truncated part of the lines. An active system of servomotors could be configured to provide an exact match of the mooring at the truncation point using force feedback and position control.

Passive systems are generally preferred as they provide simplicity, lower cost, and higher reliability. These systems utilize weights and springs to approximate the response of the truncated portion of the riser/mooring.

## 7.6 Hybrid test method

### 7.6.1 Hybrid method

Because of the basin dimension limit, the model tests cannot be used on their own anymore, and numerical analysis is needed to design the truncated (equivalent) mooring system, and numerical analysis is needed to extend the model test results to the full depth system. The combination of using the model tests and numerical analysis tools is defined as the hybrid test method.

### 7.6.2 Basic principle

Hybrid model testing addresses the practical difficulties of achieving global verification of complex deepwater floating systems. It is based on the use of validated numerical models to assist the design and interpretation of scale model tests.

The "model-of-the-model" is initially used to design the equivalent or truncated mooring system and to ensure it will function as needed to satisfy the model test objectives. As the model tests proceed, the initial model-of-the-model is updated to achieve systematic verification of agreement between numerical and physical models. This naturally leads to improved quality of the model tests as a whole. At the conclusion of the test program, the model-of-the-model is used to make rational adjustments to extrapolate system responses to full scale and to calibrate design recipes for final sizing, as needed. The process of the hybrid test method is illustrated in Fig. 7.9.

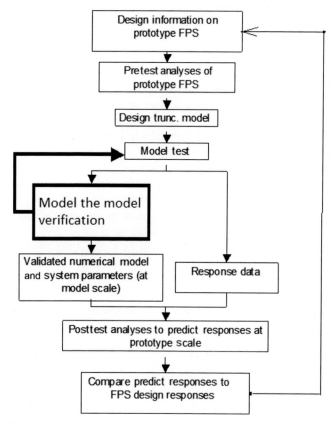

**FIGURE 7.9** Hybrid approach of model tests and numerical analysis.

### 7.6.3 Numerical tools

The successful application of the hybrid method assumes that a validated numerical model of coupled floater/mooring line/riser dynamics is available for use in the design of the truncated (equivalent) mooring system and in the extrapolation of model test results to prototype scale. The successful application of the hybrid method depends upon the following conditions:

- The coupled numerical model is able to reproduce all measured floater/mooring/riser responses of the scale model using the results of static offset tests, decay tests, wind/current force calibration tests, and regular wave tests.
- The coupled numerical model is able to reproduce all measured wave frequency (WF) floater/mooring/riser responses of the scale model using the results of random wave tests, including white noise tests and tests with superposed wind and current.

## 7.7 Model test execution

### 7.7.1 Model preparation

Prior to the model test program, extensive engineering analysis would be required to define the model test conventions, model test scale, design of the truncation system if required, and the selection of critical test cases. A technical document called model test specifications will be finalized and delivered to the model test basin.

The physical model will be constructed that includes the floater hull with its major accessories such as bilge keels, mooring fairleads, mooring and riser components, turret system, and representative topsides. The constructed physical model will be extensively calibrated in terms of dimension, weight, center of gravity, radius of gyration, etc., before being loaded into the model test basin. The physical model has to be constructed strictly according to scale and the margin of error is usually limited to $\pm 3\%$.

The constructed moorings and risers will be fixed at the basin floor and hooked to the floater. Afterward, the complete system will be tuned to match the test requirements. To verify the correctness of mooring system modeling, the restoring force curves (i.e., load vs offset curves) of the mooring system should be created and compared with the theoretical results.

### 7.7.2 Environment calibration

Prior to the actual test of designated test cases, the environmental conditions should be pregenerated and verified with theoretical values.

Irregular long-crested waves should be generated to reproduce the wave spectrum defined by significant wave height Hs, peak period Tp, and peak enhancement factor $\gamma$. For calibration purpose, wave heights shall be measured at the location where the floater at rest over the full duration of wave calibration tests.

Wind shall be generated by using a fan battery mounted in front of the model set-up. The speed of fans is adjusted so that the wind loads are correct at a vertical position of 10 m above the water surface. Current shall be generated in such a way that the current field is homogeneous (both in time and space) at any point likely to be met by the floater and mooring models.

### 7.7.3 Data collection and processing

To characterize the hydrodynamic performance, the critical response parameters are typically:

- Floater motions [from high frequency to low frequency (LF)].
- Mooring line tensions.
- Riser tension and/or minimum bending radius.

- Possible interaction (collision) related to moorings/risers/hull.
- Relative wave motions (green water, column wave run-up, air-gap).
- Global structural loads.
- Local wave impact loads, slamming.

Video cameras and underwater cameras will be mounted to continuously videotape the complete model testing process.

The total signal has to be filtered to separate WF and LF responses. During the dynamic tests, the measured values shall be derived for each signal based on the middle 3-hour long window:

- The mean.
- The standard deviation.
- The maximum and minimum.
- The most probable extreme (MPE).
- Parameters of the distribution used to derive MPE.
- The zero-crossing period.

## 7.8 Questions

1. You are designing a floating production system in a region where there is no ice condition. What types of model tests will you conduct for your project?
2. Briefly explain the law of similarity in the context of model tests.
3. What are the potential problems in a model test, when using a very small model (i.e., when choosing a scale factor that is very large)?
4. Why do we sometimes have to use a truncated mooring model rather than a full-depth mooring model?

## References

[1] B. Buchner, Numerical simulation and model test requirements for deep water development, in: Deep and Ultra Deep Water Offshore Technology Conference, Newcastle, March 1999.
[2] C. Stansberg, S. Karlsen, E. Ward, J. Wichers, M. Irani, Model testing for ultradeep waters, OTC 16587, Houston, TX, 2003.
[3] E. Ward, V. Hansen, Model-the-model: validating analysis models for deepwater structures with model tests, OTC 15350, Houston, TX, 2003.
[4] O. Waals, R. Van Dijk, Truncation methods for deep water mooring systems for a catenary moored FPSO and a semi taut moored semi submersible, DOT, New Orleans, LA, 2004.

# Chapter 8

# Anchor selection

## Chapter Outline

8.1 **Overview** 155
  8.1.1 Available anchor types 155
  8.1.2 Anchor design
    considerations 157
  8.1.3 Soil characterization 157
8.2 **Suction piles** 158
  8.2.1 Holding capacity of suction
    piles 159
  8.2.2 Suction pile installation 160
8.3 **Driven piles** 161
  8.3.1 Holding capacity of driven
    piles 162
  8.3.2 Driven pile installation 163
8.4 **Drag embedment anchors** 163
  8.4.1 Advantages and limitations
    of drag embedment
    anchors 164
  8.4.2 Holding capacity of drag
    embedment anchors 165
  8.4.3 Drag embedment anchor
    installation and recovery 165

8.5 **Vertically loaded anchors** 166
  8.5.1 Vertically loaded anchor
    for permanent and temporary
    moorings 166
  8.5.2 Holding capacity of
    vertically loaded anchors 167
  8.5.3 Vertically loaded anchor
    installation 168
8.6 **Suction embedded plate anchors** 168
  8.6.1 Advantages and limitations
    of suction embedded plate
    anchor 168
  8.6.2 Suction embedded plate
    anchor installation 169
8.7 **Gravity installed anchors** 170
  8.7.1 Torpedo anchor 170
  8.7.2 OMNI-Max anchor 171
8.8 **Questions** 173
**References** 173

## 8.1 Overview

### 8.1.1 Available anchor types

Moored floating structures impose a variety of loading conditions on the anchor system. These loads range from the horizontal load for a catenary mooring line, the combination of the horizontal and vertical load for a semi-taut and taut leg mooring line, to the vertical uplift load for a tension-leg platform (TLP). Also, soil conditions for the anchors can vary in type and properties, such as soft clay, stiff clay, sand, gravel, etc. Comprehensive engineering analysis is required to select the anchor type and to design the

Mooring System Engineering for Offshore Structures. DOI: https://doi.org/10.1016/B978-0-12-818551-3.00008-9

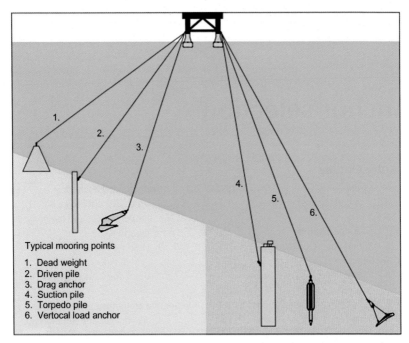

**FIGURE 8.1** Different anchor types for shallow and deep waters. *Courtesy: Vryhof Anchor.*

anchor configuration based on the loading and soil conditions. The common anchor types used in the offshore industry are as follows [1] (see Fig. 8.1):

1. Dead weight anchor
2. Driven pile anchor
3. Drag embedment anchor (DEA)
4. Suction pile anchor
5. Gravity installed anchor, including torpedo pile and others
6. Vertically loaded anchor (VLA), including suction embedded plate anchor (SEPLA)

For permanently moored floating production units, the suction pile and driven pile anchors are most commonly used [2], although high-efficiency DEA and VLA have been used for small floating units under mild environment. The torpedo pile, a gravity-installed anchor, is a relatively new anchor concept, which has been used in mobile offshore drilling unit (MODU) and permanent moorings. The anchor design requires the full range of geotechnical analysis including site investigation, soil characterization, foundation installation analysis, and foundation capacity assessment.

For drilling operations with MODUs, the DEAs are most commonly used, which can be deployed and retrieved without a specialized facility. By adjusting the fluke angle, they can work in various soil conditions such as soft or stiff clay and sand. However, they have limited capability to withstand vertical loads. VLAs are sometimes used where high vertical loads at the anchor are present. For cases with very high vertical load or where anchor movement is strictly prohibited, driven pile or suction pile can be used. Because of the temporary nature of the operation, a thorough soil investigation is normally not conducted, unless the operation is close to other facilities.

Penetration into the seabed is not possible in certain situations, such as rock seabed. It makes the anchor types described above impractical. In such cases, dead weight anchors may be considered. Reinforced concrete or scrap steel may be used for this purpose. The vertical capacity of a dead weight anchor is simply its submerged weight so the load capacity for the system is relatively small, in the order of a few tons. The horizontal load capacity can be calculated by the submerged weight multiplied by an appropriate friction coefficient. However, the dead weight anchor is rarely used for mooring operations because of the limited holding capacity.

### 8.1.2 Anchor design considerations

There are two primary design considerations for anchors at a given site.

Structural design—The anchor design is evaluated under the following loads:

- maximum loads imposed by the anchor line,
- maximum loads imposed during transportation and installation, and
- fatigue damage sustained over the lifetime.

Geotechnical design—The anchor design is evaluated by the following analyses:

- soil–structure interaction analysis to determine soil reactions acting on the anchor for input into the structural design of the anchor,
- geotechnical holding capacity analysis to determine the required size and depth of embedment of the anchor to achieve the desired capacity, and
- checking the ability of the anchor to achieve the required embedment and ease of removal.

The anchor failure mechanism in the soil depends on various factors, including anchor geometry, the load inclination, the depth of the load attachment point, and the soil shear strength profile. For pile anchors, the loading point (padeye location) is usually positioned well below the mudline to get the optimal holding capacity.

### 8.1.3 Soil characterization

Before the engineering and installation phases of a project, a site investigation for the field is usually conducted. It provides essential information, such as subsea terrain, topography (bathymetry), soil properties, etc., for determining the anchor locations and sizing of the anchors. Some MODU operations, however, may simply use most available information without conducting a site investigation. A site investigation campaign includes geophysical and geotechnical surveys that may cover a subsea survey, positioning of facilities, soil sampling, and soil tests. More discussion on site investigation is presented in Section 11.1.

The soil type is classified mainly by grain size distribution. In general, the soil types encountered in anchor design are sand and clay, with grain diameter from 0.1 μm to 2 mm. However, mooring locations consisting of soils with grain sizes above 2 mm, such as gravel, cobbles, boulders, rock, etc. also exist.

Soil strength is generally expressed in terms of shear strength parameters of the soil. One of the key soil parameters in the design of an anchor is the undrained shear strength. Clay type soils are generally characterized by the undrained shear strength, together with the submerged unit weight, the water content and the plasticity parameters. The undrained shear strength values are usually measured in the laboratory. On site, the values can be estimated from the results of Cone Penetration Tests and vane tests. The mechanical resistance of sandy soils is predominantly characterized by the submerged unit weight and the angle of internal friction. These parameters are established in the laboratory.

Typical deepwater soil deposits consist of soft clay with occasional sand layers. Over the past thousands of years, sea levels have varied by some 300 ft. or more. In nearshore areas, particularly near river mouths, deposition can be relatively rapid and the soil accumulates faster than the pore water can escape. This condition leads to very weak or so-called under consolidated conditions. In deep water or away from sediment sources, the accumulation rate can be extremely slow (millimeters per thousand years) such that the pore pressures remain at hydrostatic values during the deposition process. This gives rise to normally consolidated clays. The clay shear strength in these soils will increase more or less linearly with depth.

## 8.2 Suction piles

Suction piles are cylindrical anchors of a large diameter, typically from 4 to 6 m (as shown in Fig. 8.2). They are installed partially by self-weight penetration, with penetration to full installation depth accomplished by the application of "suction," which is actually a differential pressure induced by pumping through a valve in the top cap. Suction piles have advantages over driven piles mainly because they don't require heavy underwater hammers. The pile aspect ratio (ratio of length over diameter) is largely dictated by

**FIGURE 8.2**    Suction piles. *Courtesy: SPT Offshore.*

considerations for safely installing the pile. Optimal aspect ratios are typically in the following ranges [3]:

- For dense sands—less than 1.5.
- For stiff clays—between 1.5 and 3.
- For soft clays—greater than 5.

Suction piles are normally designed with relatively thin walls; typical diameter-to-wall-thickness ratios are in the range of 125−160, in contrast to 10−40 for driven piles [3]. Various internal plate stiffeners and ring stiffeners are required for preventing structural buckling during installation and structural failure during operation. This type of anchor has been used for direct vertical loading (e.g., TLP), but is mostly used for catenary and taut mooring systems.

## 8.2.1    Holding capacity of suction piles

There are three categories of analysis tools that can be used to determine the holding capacity of suction piles [4]:

- Limit equilibrium or plastic limit analysis methods − models involving soil failure mechanisms.
- Semiempirical methods—highly simplified models of soil resistance including beam-column models. Note that these methods, such as beam column models, are not preferred because suction piles have a different failure mechanism from driven piles.
- Finite Element Analysis (FEA)—advanced numerical analysis.

For deepwater permanent moorings, the design focus is on the ultimate capacity of the suction pile and not on the load deflection behavior. FEA or limit equilibrium techniques and plastic limit analyses calibrated to the FEA model are recommended. For temporary catenary mooring systems, where loads are mainly horizontal, semiempirical methods such as beam-column analyses may be used if other methods are not available.

The depth of the mooring line attachment point (padeye location) can have a substantial effect on the holding capacity. The maximum capacity will vary depending on the aspect ratio of the pile, soil strength profile, and other soil parameters. Attachment at the optimum point (maximum capacity) results in just translation with no rotation of the pile at failure load. The optimum point is generally located about two-thirds of the way down from the pile top if soils consist of normally consolidated clays, such as the typical soils in the Gulf of Mexico.

## 8.2.2   Suction pile installation

Initially, the suction pile is vented at the top to allow water to escape during dead weight penetration. The penetration is resisted by soil shear stresses acting on the walls of the pile (external and internal) and by bearing resistance acting at the tips. Once the pile comes to rest under its weight, the top is sealed, and the pressure inside the pile is slowly lowered by an remotely operated vehicle (ROV). In simple words, seawater is pumped out to create negative pressure (i.e., suction). Clay soils are relatively impermeable, so the pressure gradient through the soil causes negligible flow. This pressure difference forces the pile into the soil until it reaches the target penetration depth (Fig. 8.3). Careful monitoring is carried out during installation to ensure that the soil plug does not fail and heave upward and that lowering the pressure does not cause buckling of the pile. Either occurrence would require removal of the pile. Measures must be taken to ensure that the orientation and inclination remain within tolerance. The pile can be retrieved by pumping water back inside the pile.

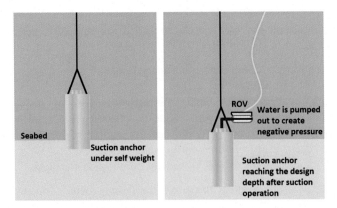

**FIGURE 8.3**   Mechanism of pile penetration by suction pressure.

Suction anchors can be placed accurately with respect to horizontal position, verticality, and orientation. Installation is closely monitored and controlled. Data recorded during installation serves to either verify the design assumptions or to allow the capacities to be revised. Proof load testing to verify the holding capacity is not normally required for suction or driven piles, since their positioning is well controlled, and their design is supported by analyses based on site-specific soil data.

The prediction models for installation are relatively simple and similar to conventional pile design. The soil shear resistance is estimated using the remolded shear strength of the soil and the tip resistance is estimated from conventional bearing capacity theory. At any depth, the total soil resistance is estimated to determine the resistance that must be overcome by a combination of the pile weight and negative pressure. The biggest uncertainty in this prediction has been found to be in estimating the resistance of internal stiffeners. The prediction model provides the installation team with expected performance including likely ranges of uncertainty.

## 8.3    Driven piles

Driven piles are large diameter (1−3 m) open-ended pipes, which are relatively simple to fabricate (as shown in Fig. 8.4). They can be installed in a

**FIGURE 8.4**    Driven piles. *Courtesy: InterMoor.*

wide range of soil conditions. They can be driven 100 m or more into the seabed where they can mobilize the high soil strengths at these depths and individually develop very large uplift resistances, for example, 5000 tons or more. Thus, in principle, they are well suited for TLP support (direct uplift) compared to alternatives. They can also be used for horizontal or inclined loads. They are, therefore, also a good option for a taut or catenary mooring system. Because pile hammer-driving and handling become more difficult in deeper water, driven piles are less common for ultradeep water. Nevertheless, this method has been used in water depths of up to 2400 m.

### 8.3.1 Holding capacity of driven piles

The key element of driven pile design for uplift loading is the pile's axial capacity. Two cases can be considered for estimating axial pile capacity. In one case, the soil plug (the soil trapped inside the pile) is assumed to remain in place while the pile moves up in "cookie cutter" style around the plug and through the soil. In this model, soil shear resistance develops on the outside (external skin friction) and inside (internal skin friction) as well as reverse bearing resistance on the pile annulus. For calculation purposes, the unit internal and external skin friction values are typically assumed to be equal. In the other case, the soil plug is assumed to move up with the pile such that a reverse bearing failure develops across the whole pile tip. In this case, which generally governs for long piles, only external skin friction is mobilized. The capacity of the pile is computed for both assumptions and the minimum value is taken for design. In normally consolidated clays, the uplift resistance of the tip is a small fraction of the external shaft resistance such that it is often ignored. Of course, the detailed calculations depend on the soil type and design parameters.

Driven piles can also be used as anchors to resist lateral loads or inclined loads, although they are generally not as effective in this role. One of the earliest attempts to model laterally loaded piles was to idealize the soil as a bed of linear springs. In such a model, there is no coupling of soil resistance from point to point along the pile, that is, the soil resistance at any point on the pile is simply proportional to the displacement of that point. Although this behavior is clearly oversimplified, the model does seem to capture the basic physics of the system, is surprisingly robust, and is widely used.

The beam-column method is more commonly used today, which is semiempirical with soil represented as uncoupled nonlinear springs along the pile boundary. Soil springs are based on full-scale lateral load tests, and soil behavior is characterized and correlated with measurable soil properties such as soil type and strength parameters [5].

## 8.3.2    Driven pile installation

Onshore methods for driving piles include impact hammers powered by steam, air, or diesel. Historically, onshore technology was adapted to offshore pile installations using a "follower" that was attached to the pile to transmit the stress wave from the water surface to the pile. This approach was technically feasible to water depths up to approximately 300 m. In the mid-1970s, development at greater water depths spurred the development of underwater hydraulic hammers (shown as a yellow cylindrical tool in Fig. 8.4). The development of underwater pile driving hammers has made it possible to drive piles without a follower section to the surface. This has made driving piles in deep water (up to 2400 m) feasible today. A variety of hammers are available. and large barges are typically required for transportation and installation.

## 8.4    Drag embedment anchors

A DEA is a bearing plate, known as the fluke, inserted into the seabed by dragging it with wire rope or chain (Fig. 8.5). The fluke is attached to the

**FIGURE 8.5**    Typical DEA configurations. *DEA*, Drag embedment anchor. *Courtesy: Vryhof Anchor and Mooreast.*

anchor line by a "shank" comprising one or more plates. Self-embedment of the anchor is achieved by controlling the line of action of the mooring line force, by setting the fluke-shank angle such that "soil failure" occurs roughly parallel to the fluke, so that the anchor will move downward when dragged. DEAs are normally designed for several possible fluke-shank angle settings according to the soil type. In stiff clays and sands, the fluke-shank angle is typically set around 30 degrees, while 50 degrees is typical for soft clays.

Traditional DEAs were initially used for mobile mooring operations. DEA technology has advanced considerably, and the new generation of fixed fluke DEAs can develop a much higher holding capacity under various soft soil conditions. High efficiency DEAs are generally considered an attractive option for mooring applications because of ease of installation and proven performance. The anchor section of a mooring line can be pre-installed and proof loaded prior to arrival of the floating vessel on location.

### 8.4.1    Advantages and limitations of drag embedment anchors

DEAs are less expensive to install compared to suction and driven piles, and they are generally more efficient than suction and driven piles in terms of the ratio of load capacity to anchor weight [3]. For example, a high-efficiency DEA can potentially have a holding capacity of up to 20−90 times its own weight depending on the soil condition. However, DEAs do not have the ability to reach a precise position like suction and driven piles. Further, the load capacity of the anchor depends on the depth of anchor penetration, which cannot be predicted with a high degree of certainty. Nevertheless, the uncertainty in DEA load capacity can be mitigated substantially by proof load testing of anchors following installation, which is usually required for MODU and permanent moorings. It should be noted that the proof loading of DEAs will require on board tensioning equipment or anchor handling vessels (AHVs) with certain bollard pull capability.

DEAs in sands and stiff clays experience minimal penetration into the seabed, typically less than 1−2 fluke lengths. Consequently, the vertical load capacity is minimal, and their application is restricted to catenary systems. Effectively, the anchor provides resistance to horizontal loads, while the dead weight of the anchor and chain resists vertical loads. In contrast to the case of sands and stiff clays, DEAs in soft clays will penetrate to substantial depths, in the order of tens of feet in some cases. Since soft clay profiles typically exhibit increasing strength with depth, increased penetration leads to increased DEA load capacity, including a substantial capability for resisting vertical loads. Thus, DEAs in soft clay can provide sufficient anchorage for both catenary and taut mooring systems.

DEAs can also penetrate hard soils, cemented layers, and soft rocks (chalk, calcarenite, corals, limestone, etc.), if designed properly. Such proper design typically requires sufficient structural strength to sustain extreme concentrated loads, a serrated shank and cutter-teeth (pick points) for better penetration.

## 8.4.2   Holding capacity of drag embedment anchors

Among the simplest drag anchor capacity prediction methods are charts which provide estimates of holding capacity, drag distance, and penetration depth as a function of anchor weight for a range of soil types. These charts are usually anchor specific, based on full-scale or model testing and field experience. Note that the typical UHC (Ultimate Holding Capacity) charts presented by suppliers are not meant for use as design guidelines, but as a rough guidance for estimating the likely anchor size. The UHC charts are only applicable for homogenous generic soil types of unlimited thickness. For correct anchor type/size selection and for determination of anchor parameters, anchor manufacturers should be contacted and be given the available site/soil data, the mooring design loads, and the anchor line particulars.

Uncertainties are significant because of limited test data, especially data for large anchors in various soil conditions. From a mechanics perspective, the anchor weight itself plays only a minor role in capacity development. The more important factor is the fluke area which, of course, is correlated with anchor weight.

Analytical tools based on limit equilibrium principles for anchor embedment and capacity calculation in soft clay are now available [6,7]. These tools allow modeling of different anchor designs and provide detailed anchor performance information such as anchor movement trajectory, anchor rotation, mooring line profile below seafloor, and ultimate anchor capacity, etc. However, there are certain requirements for these tools to yield reliable predictions.

## 8.4.3   Drag embedment anchor installation and recovery

Discussion on DEA installation can be found in Section 11.3 "Deployment & Retrieval of Temporary Mooring."

The anchor recovery load will depend on (1) soil characteristics, (2) anchor size and embedment depth, (3) applied installation load or the highest tension the anchor is subjected to, and (4) recovery load application angle. In general, the anchor recovery loads in soft cohesive soils (clays, silts) are higher than the recovery loads in cohesionless soils (sands, gravels). The recovery loads in cohesionless soils are in the order of 20%−30% of the installation load or of the highest tension that the anchor is subjected to.

The thixotropy and consolidation of the clay are important characteristics of clays that would dictate the post installation capacity of the anchor (known as setup or consolidation effects). As a rule of thumb, the recovery loads in clays will be in the range of 80%−100% of the anchor installation load or of the highest load that the anchor is subjected to during its operation. For clays of high sensitivity, the recovery loads may go above these ranges to 110%−140%.

When recovering anchors embedded in clays, patience should be practiced as the recovery may take time. It is better to gradually increase the tension to the estimated recovery loads and hold this tension for a certain time (say, 10−30 minutes). This will allow to overcome the suction forces and gradual rotation of the anchor. Following the dissipation of suction effects and anchor rotation, the anchor will break out easily. Anchor sizing for MODU application should consider both UHC requirements and Recovery loads.

## 8.5 Vertically loaded anchors

To enhance the pullout resistance of DEAs in soft clays, VLAs were developed. A VLA is installed in the same manner as a DEA, but it features a releasable shank that can be opened after drag installation. Most VLA designs employ a shear pin that ruptures when the mooring line force exceeds a certain level, although more recently some mechanical release mechanisms have been developed. VLAs are suitable for anchoring in soft clay or layered soil consisting of soft clays. The use of VLAs in sand and stiff clay is not recommended. At present, two brands of VLAs are available, Stevmanta and Dennla (as shown in Fig. 8.6), which are manufactured by Vryhof and Bruce, respectively [8,9].

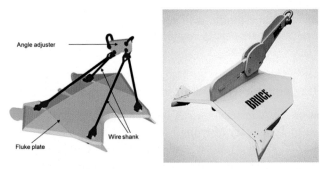

**FIGURE 8.6** Examples of VLAs, Stevmanta (left) and Dennla (right). *VLAs*, Vertically loaded anchors. *Courtesy: Vryhof Anchor and Bruce Anchor.*

### 8.5.1 Vertically loaded anchor for permanent and temporary moorings

The permanent VLAs typically have two operating modes: an installation mode and normal loading mode. In the installation mode, the load arrives at an angle of approximately 40−60 degrees with respect to the fluke. After triggering the anchor, either by shearing a shear pin or switching from the installation line to the operating line, the load becomes perpendicular (90 degrees) to the fluke. Depending on anchor and mooring line dimensions and the sensitivity of the soil, this change in load direction may generate a holding capacity of 1.5−2.5 times the installation load [6]. With the anchor shank perpendicular to the fluke, the VLA will be pulled out when overloaded, and therefore requires a higher factor of safety than conventional DEAs [4].

Both Vryhof and Bruce have offered an alternative version of VLA to their customers. In this version, after shearing a shear pin and triggering the anchor, the load becomes nearly perpendicular (near-normal) to the fluke, and the VLA will continue to drag deeper instead of being pulled out when overloaded. Since this VLA behavior is similar to that of a drag anchor, the factor of safety for a DEA can be used. The near-normal setting can be used for permanent and MODU moorings.

The version of the VLA for permanent moorings does not have a reverse loading release mechanism, and the anchor is recovered by pulling backwards on a second line. Consequently, the permanent version of the VLAs can resist reverse loading and therefore may also have limited resistance to out-of-plane (sideway) loading. However, care should be taken when using VLAs designed for near-normal loading as they may lose capacity if rotated approximately 90 degrees in the vertical plane after a windward mooring line failure and leeward line direction change as the MODU drifts off location over a leeward anchor [10].

Both Vryhof and Bruce have developed VLAs that are suitable for MODU (i.e., temporary) moorings as they are easier to deploy and recover. These VLAs are designed to be retrieved by loading in the reverse direction to operate a release mechanism, permitting recovery of the anchor by the mooring line. As an example, the Stevmanta VLA can be equipped with an optional recovery system. The recovery system consists of two special sockets which connect the front wires to the fluke. To recover the anchor, the mooring line is pulled backwards, and the front sockets will disconnect from the fluke. The Stevmanta VLA is now pulled out of the soil using just the rear wires. These anchors can be deployed from the MODU or preset. The fluke setting for these anchors is usually near-normal, but a normal fluke is also possible.

## 8.5.2   Holding capacity of vertically loaded anchors

Among the simplest VLA capacity prediction methods are charts which provide estimates of ultimate pull-out capacity and installation load as a function of fluke area and mooring line diameter. These charts provide rough estimates, and uncertainties are significant.

There are two design codes addressing VLA design and site assessment. API RP 2SK [4] provides factors of safety, general guidance, discussion, and references for the geotechnical analysis of VLAs in the installation and operating modes. A more detailed discussion on performance prediction methods for VLAs can be found in API RP 2SK. DNV-RP-E302 [6] provides detailed guidance and equations for the analysis of plate anchors in both the installation and operating modes.

The holding capacity of a VLA depends strongly on its final orientation and depth below the seabed, hence the prediction of the anchor trajectory during installation is critical. Methods for predicting the installation performance are either empirical (based on correlations with observed anchor performance) or are based on geotechnical analysis of the anchor system and installation scenario.

### 8.5.3 Vertically loaded anchor installation

The penetration depth of a VLA determines its holding capacity. The ability to track the position of a VLA during installation and to know the final position of a VLA after installation is important for permanent moorings. Both Vryhof and Bruce have developed tracking devices for their VLA anchors. In addition to the uncertainty of the final position of the anchors, there is the uncertainty of the drag distance required to embed the anchor to obtain the desired holding capacity.

## 8.6 Suction embedded plate anchors

The SEPLA, developed by Dove et al. [11], is a plate anchor installed by attaching it to the tip of a suction pile called the suction follower [12]. The suction follower is installed in the conventional manner but retracted by overpressure to leave the SEPLA behind. The vertically oriented plate is then "keyed" to turn perpendicular to the direction of the applied mooring line load (Fig. 8.7). Some loss of embedment may occur during the keying process, causing a reduction in load capacity. Some SEPLA design concepts incorporate a flap designed to minimize the loss of embedment during keying. However, it has been shown by later studies [3] that the flap may hinder the keying of the plate and increase the loss of embedment.

### 8.6.1 Advantages and limitations of suction embedded plate anchor

SEPLAs have the benefit that they combine the precise vertical and horizontal positioning of a suction pile with the lightness and efficiency of a plate anchor. Deployment is largely limited to soft clay soil profiles [3]. Since the anchor is deeply embedded, it has capabilities for resisting vertical loads and is a suitable anchor alternative for taut mooring systems. Installation time is slightly longer, in the order of 20% greater than for a conventional suction pile installation [3], since both suction follower installation and overpressure

**FIGURE 8.7**   A SEPLA is installed with the aid of a suction follower. *SEPLA*, Suction embedded plate anchor. *Courtesy: InterMoor.*

retraction are required. Accordingly, installation costs are greater than that for VLAs, and the benefit of precise positioning of the plate is obtained at the price of higher installation costs.

Suction piles require large transport vessels due to their large sizes, which can drive the cost higher. By contrast, SEPLA deployments require the transport of only a single suction follower in conjunction with a large number of compact plate anchors that can easily be arranged on the deck of a transport vessel. Overall, for soft clay soil profiles, SEPLAs can be a competitive alternative, when capabilities for resisting vertical/inclined loads and precise anchor positioning are required.

### 8.6.2   Suction embedded plate anchor installation

The SEPLA has an advantage over DEAs, because its penetration depth and position are known. Since the anchor falls within the guidelines required for DEAs, the keying load or proof load must also reach 80% of the design load during installation. The holding capacity is therefore limited to the bollard pull of the installation vessels. The final orientation of the anchor after preloading is assumed to be normal to the anchor line load; however, complete reorientation of the plate cannot be assured and thus is an uncertainty associated with the installation of a SEPLA.

## 8.7   Gravity installed anchors

Gravity installed anchors (also known as drop anchors) penetrate into the seabed by the kinetic energy obtained from free fall through the water column by their own weight. This installation method allows for reduced overall installation time and does not require installation vessels with significant bollard pull. This type of anchor typically is cylindrical with a nose cone, with up to four stabilizing fins at the top of the pile. Examples of drop anchors include torpedo anchor, OMNI-MAX anchor, Deep Penetrating Anchor, and others.

### 8.7.1   Torpedo anchor

The torpedo anchor (Fig. 8.8) is an innovative anchor concept developed by Petrobras. As an example, a typical torpedo anchor has the following parameters for permanent mooring systems:

- Length ∼20 m, and diameter ∼1 m
- Weight ∼100 mT in air; weight of the latest design ∼120 mT
- Drop height ∼100 m above seabed with the full mooring line attached
- Terminal velocity in the range of 30−50 m/s.

Torpedo anchors offer the advantage of quick, economical installation. They can be installed with AHVs and require minimal mechanical equipment, in contrast to the heavy underwater hammers required for piles or pumps for suction piles. They are typically installed in soft to medium clay conditions. Since they embed relatively deep, torpedo anchors can resist both horizontal and vertical loads and are thus suitable for both catenary and taut mooring systems.

Two vessels are normally required for installation, one vessel with an A-frame to handle the anchor using a deployment line, and another vessel to handle the mooring line. However, Petrobras has used only one AHV to install the torpedo anchors. An ROV is used to assist the installation and survey the as-installed position of the anchor. The anchors can be positioned accurately, and the penetration of the anchors can be determined after installation by using an ROV to observe the penetration markings on the mooring line. No bollard pull is required to key the anchors. Orientation of the anchor after installation is not a concern, since the padeye is located at the top of the anchor, and the padeye design allows for a mooring line load in any direction. Overall, the installation procedure is simple. Petrobras has installed many torpedo anchors offshore Brazil for permanent mooring systems.

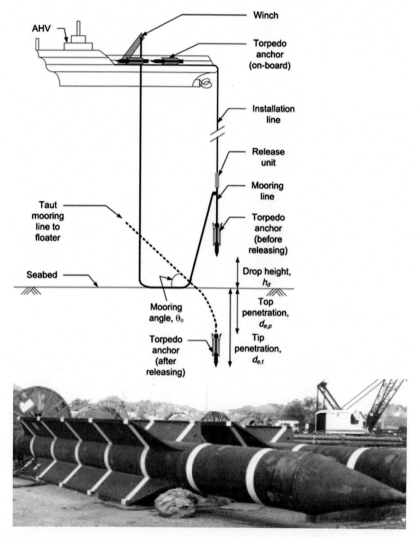

**FIGURE 8.8**   Torpedo anchor and installation by gravity. *Courtesy: InterMoor.*

## 8.7.2   OMNI-Max anchor

The OMNI-Max anchor, developed by Delmar Systems, provides another anchor alternative [13]. It can be installed where a typical mooring suction pile can be installed and is intended to offer a lower cost alternative to the suction pile. The anchor concept is illustrated in Fig. 8.9.

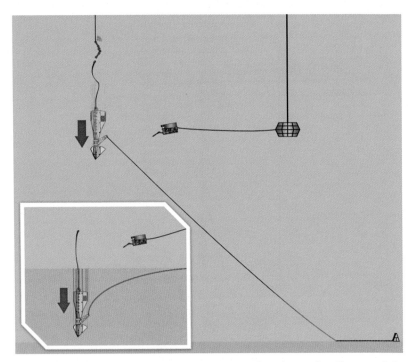

**FIGURE 8.9**   An OMNI-Max anchor penetrates the seafloor by gravity. *Courtesy: Delmar Systems, Inc.*

The anchor is first lowered from an AHV to a predesignated height. Once the anchor is at the correct elevation from the seabed, a remote release actuator is activated, and the anchor plunges down into the seabed. After being pulled by the mooring line, the anchor rotates and penetrates deeper into the soil providing the desired vertical and horizontal holding capacity [14].

The components of the OMNI-Max anchor consist of a nose plate, a rotating loading arm, a load ring, and a tail plate. There are several distinct features of the anchor:

- It is a relatively small sized anchor. Compared to a similar capacity sized suction pile, it is approximately 50% smaller.
- It has an omnidirectional mooring attachment arm. During installation, this attribute alleviates most alignment concerns.
- With adjustable fluke fins, the anchor can be set up prior to deployment for a predictable rotational behavior in different soil profiles and loads.
- When the anchor is loaded, the initial behavior is to rotate and become more perpendicular with the load angle. Following the rotation, it will dive deeper to stronger soils until the soil resistance is equivalent to the mooring load.

- To retrieve the anchor, the vessel winch pulls the recovery line attached to the top of the anchor.

The OMNI-Max anchor had its first use in 2007 in the Gulf of Mexico for a MODU mooring. There are many installed in the Gulf of Mexico and West Africa. The installation requires one AHV, and all anchors can be transported at one time.

## 8.8    Questions

1. In Fig. 8.2, there are three wide (fat) suction piles and two slender ones. Why are they so different in shapes?
2. There is a hammer in Fig. 8.4. Could you point out where it is?
3. Why are DEAs so widely used? Name at least two advantages of them.
4. You are designing a catenary mooring system for a floating wind turbine. Which type of anchor would you select if the cost is the main decision criterion?
5. You are designing a taut leg system for a tender assisted drilling (TAD) that will be moored in 400-m water next to a compliant tower for several years. To keep the gangway bridge connected and operable even in a 10-year storm, the TAD's offset must be minimized with a very taut mooring system. Which type(s) of anchor would you recommend, assuming the soil is soft clay? Why?

## References

[1]   Vryhof Anchors, Vryhof Manual, the Guide to Anchoring, Vryhof Anchors B.V, 2015.
[2]   C. Ehlers, A.G. Young, J.-H. Chen, Technology assessment of deepwater anchors, in: Offshore Technology Conference. Offshore Technology Conference, 2004.
[3]   C. Aubeny, Geomechanics of Marine Anchors, first ed., CRC Press, 2018.
[4]   API, Design and analysis of stationkeeping systems for floating structures, API Recommended Practice 2SK, third ed., 2005.
[5]   API, Recommended practice for planning, designing, and constructing fixed offshore platforms—working stress design, API RP2A-WSD, 22nd ed., 2014.
[6]   Det Norske Veritas, Recommended practice, design and installation of plate anchors in clay, DNV-E302 Det Norske Veritas, 2013.
[7]   R. Dahlberg, DNV design procedures for deepwater anchors in clay, OTC 8837, Houston, TX, 1998.
[8]   R. Ruinen, G. Degenkamp, Vryhof Anchors B.V., First application of 12 Stevmanta anchors (DREPLA) in the P27 taut leg mooring system, Proceedings of 11th DOT Conference (Deep Offshore Technology), Stavanger, Norway, 19–21 October 1999.
[9]   P. Foxton, Bruce Anchor Limited, Latest development for vertically loaded anchors, in: IBC 2nd Annual Conference—Mooring and Anchoring, Aberdeen, 1997.
[10]  API, Gulf of Mexico MODU mooring practices for the 2007 hurricane season – interim recommendations, API RP-95F, second ed., 2007.

[11] P. Dove, H. Treu, B. Wilde, Suction embedded plate anchor (SEPLA): a new anchoring solution for ultra-deepwater mooring, in: Proceedings of the Deep Offshore Technology Conference, New Orleans, LA, 1998.

[12] B. Wilde, H. Treu, T. Fulton, Field testing of suction embedded plate anchors, in: Proceedings, ISOPE, 2001.

[13] E. Zimmerman, M. Smith, J.T. Shelton, Efficient gravity installed anchor for deepwater mooring, in: Offshore Technology Conference. Offshore Technology Conference, 2009.

[14] J. Liu, C. Han, L. Yu, Experimental investigation of the keying process of OMNI-Max anchor, Mar. Georesour. Geotechnol. (2018) 1−17. Available from: https://doi.org/10.1080/1064119X.2018.1434841.

# Chapter 9

# Hardware—off-vessel components

## Chapter Outline

9.1 **Mooring line compositions** 175
9.2 **Chain** 176
  9.2.1 Studlink versus studless 177
  9.2.2 Chain grades 177
  9.2.3 Manufacturing process 178
9.3 **Wire rope** 180
  9.3.1 Six-strand versus spiral strand 181
  9.3.2 Corrosion protection 182
  9.3.3 Termination with sockets 183
9.4 **Polyester rope** 183
  9.4.1 First use of polyester mooring in deepwater 185
  9.4.2 Rope constructions 185
  9.4.3 Polyester stretch 187
9.5 **Other synthetic ropes** 187

9.5.1 Nylon rope 188
9.5.2 High modulus polyethylene rope 188
9.5.3 Aramid rope 190
9.5.4 Considerations for moorings in ultradeep waters 190
9.6 **Connectors** 191
  9.6.1 Connectors for permanent moorings 191
  9.6.2 Connectors for temporary moorings 193
9.7 **Buoy** 195
9.8 **Clump weight** 196
9.9 **Questions** 197
**References** 197

## 9.1 Mooring line compositions

A mooring line can incorporate chain, wire rope, synthetic fiber rope, or a combination of these. In shallow water, chain is used extensively, which many refer to as an "all-chain" design. It is a simple and effective design taking advantage of the fact that chain is sturdy, has a good resistance to seabed abrasion, and provides added holding capacity to the anchor. To enhance the station-keeping performance, some mooring designers fit clump weights on ground chain near the touch down point. The additional weight can increase the restoring force of the mooring system, as the vessel would have to lift those weights before it can offset further.

In deeper water, an all-chain system may become too heavy. The weight of chain causes the catenary shape to dip (i.e., sag), and the angle at the top of a mooring line becomes steeper. The result is a less-efficient mooring system that provides a reduced restoring force to the floating vessel. On top of

Mooring System Engineering for Offshore Structures. DOI: https://doi.org/10.1016/B978-0-12-818551-3.00009-0

that, the added chain weight must be carried by the floating vessel thus reducing the vessel's payload capacity. This is where wire ropes can be utilized in a mooring system. Because of its lighter weight, wire rope alleviates the weight challenge found with all-chain designs. Meanwhile it offers a higher restoring force at the same given pretension because of the less-steep catenary shape of lighter mooring lines. As such, wire rope was introduced to the offshore mooring industry and became popular when drilling and production vessels went to deeper water. Wire rope was widely used as the middle segment in a mooring line, making it a "chain−wire−chain" design. Most designers still prefer chain over wire rope for the bottom (ground) segment, because wire rope can wear under long-term abrasion on seabed. For mobile offshore drilling unit (MODU) moorings, the design becomes simply "wire-chain," as wire ropes are typically deployed directly from winches on the MODU deck where there are no top chains. Buoys can be added to the "wire−chain" system to increase restoring force and reduce the vertical load on the vessel. However, there are problems associated with the use of buoys, which are discussed later in this chapter.

For vessels stationed in deep or ultradeep water, polyester rope has been increasingly favored over wire rope due to its much lighter weight and lower stiffness. Polyester rope is not only highly competitive in cost, but also offers longer fatigue life than wire rope. Thus "chain−polyester−chain" designs have become a standard configuration for mooring systems in ultradeep water. To reduce weight, designers continue to extend the polyester segment as much as possible by minimizing the lengths of both top and bottom chains.

## 9.2 Chain

The most common component used in mooring lines is chain, which is available in different diameters and grades. Offshore mooring chains are typically quite large in size with bar diameters ranging from 70 to 200 mm. By their appearance, two different designs of chain are used frequently, studlink and studless chain, as shown in Fig. 9.1.

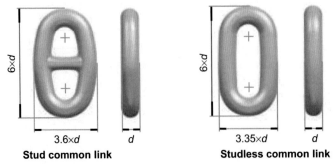

**FIGURE 9.1** Stud and studless chain links. *Courtesy: Vicinay.*

## 9.2.1    Studlink versus studless

The studlink chain is most commonly used for temporary moorings that have to be deployed and retrieved numerous times during their lifetime. A good example is the chain used on drilling semisubmersibles. By contrast, studless chain is often used for permanent moorings, such as those for F(P)SOs, catenary anchor leg mooring (CALM) buoys, spars, and production semisubmersibles. Those floating production facilities are designed to stay at a site for 20−30 years, and their mooring lines are not intended to be retrieved once installed.

Studlink chain has a stud fitted inside the oval link. Its purpose is simply to avoid tangling of chain. In the early history of shipping, chain was used to deploy and retrieve anchors, and tangling of those heavy chains was a major problem with no easy solution. The consequence could have been as severe as missing the scheduled sail away date required for on time deliveries. Therefore studlink chain was utilized to effectively mitigate such a problem. Today, studlink is still the most common chain type used by ships.

For the same reason, the offshore industry has been using studlink chain for mobile (temporary) mooring systems. MODUs normally stay at a site only for a few months. They need to retrieve and redeploy their mooring chains every time they move from one site to another. Studlink chain allows smooth handling without the problem of tangling.

In the 1990s studless chain started to gain wide acceptance in the application of permanent moorings. Studless chain is about 10% lighter than the stud chain, and still has the same breaking strength. Studless chain also offers the added advantages over studlink chain such as no loose studs, no cracks at the stud weld, and studless chain is easier to manufacture and inspect. Therefore most permanent moorings choose studless chain over studlink chain.

## 9.2.2    Chain grades

Offshore mooring chain can be obtained in several grades from R3, R3S, R4, R4S to R5 [1]. Among them, R5 has the highest strength. Fig. 9.2 shows the minimum breaking loads (MBLs) against chain diameter for studless chain. Mechanical properties of these grades can be found in ISO 20438 [1]. These grades were defined gradually by Det Norske Veritas (DNV) in 1985, 1995, and 2008 [2,3]. Five chain grades (R3, R3S, R4, R4S, and R5) are covered in DNV OS E302 "Offshore Mooring Chain" in 2008. Other class societies, such as American Bureau of Shipping, Bureau Veritas, etc., have published similar guidance notes or specifications.

It is worth noting that offshore rig quality (ORQ) chain was the predecessor to R3 chain. The chain grade was introduced to meet the demand of high strength mooring chains for MODUs. It was first defined in the American Petroleum Institute (API) Specification 2F "Mooring Chain" issued in 1974 [4].

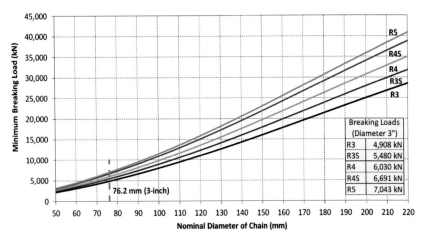

**FIGURE 9.2**    Breaking strength of chains in different grades.

The specification covered the material requirements as well as manufacturing and testing of studlink ORQ chain with a minimum tensile strength of 641 MPa. ORQ chain has mechanical properties slightly lower than those of R3 grade, and has been used in large quantities over the years with generally good performance.

It is important to note that there is another category of chain called ship anchoring chain or marine chain. They were introduced before offshore mooring chain, and have three steel grades, Grades 1, 2, and 3. Their MBL values were defined in the Anchors and Chain Cable Act by the UK Parliament in 1970 [5,6]. With lower tensile strengths, Grades 1 and 2 are not recommended for offshore mooring operations. Grade 3 has seen some very limited uses in the mooring systems of CALM buoys.

When specifying mooring chain of higher grades, the designer needs to carefully consider the fracture toughness and the likelihood of hydrogen embrittlement. All chains need to pass strict testing to ensure their strength and mechanical properties meet requirements before they are certified by a Classification Society.

For studlink chain of a lower grade such as R3, studs are often welded on the side opposite to the flash weld. Studs are normally not welded for higher grades. While stud serves a good purpose, they can become sources of problems. Common issues include loose studs, fatigue cracks, and fractures at the stud weld or stud footprint.

### 9.2.3  Manufacturing process

Offshore mooring chains are made through a complex manufacturing process. Due to the fact that each individual link is made one by one, any defect in a single link can have a detrimental impact on the reliability of the entire mooring chain. This is very different from wire rope and synthetic fiber rope. Ropes are made of wires or fibers that are weaved in a continuous process, so they have less of the reliability problem that chain may experience. For this reason, it is beneficial for mooring engineers to gain some understanding of the manufacturing process for chain. Note that reliability and integrity of all mooring components are further discussed in a separate chapter later in the book.

Manufacturing of chain begins with cutting of the steel bars. Each bar is cut to the required length. After the preheating, the bar goes to the bending machine, where it is automatically bent and joined with the previous link, as shown in Fig. 9.3. It then goes through a flash butt welding process, where

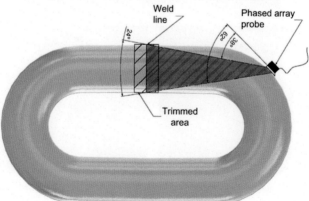

**FIGURE 9.3**   A new link is ready to be flash welded (top); PAUT can detect defects in weld zone (bottom). *PAUT*, Phased Array Ultrasonic Testing. *Courtesy: Ramnas.*

two ends are welded with no material addition. Note while flash welding is a proven technology, there is a chance that it introduces small defects in the weld zone.

After having undergone nondestructive testing, the welded chain then passes on to the heat treatment phase, which gives the material the final mechanical properties. After that, the chain is proof loaded to test its resistance to tensile loads. 100% of the links are tested in a proof load test bench. Then the chain is shot-blasted to prepare the surface for final nondestructive inspection. Each link is inspected using fluorescent magnetic particles. In addition, its weld can be inspected using Phased Array Ultrasonic Testing (PAUT). PAUT is an effective method to find defects in flash welds, and therefore it is specified more and more by clients. Fig. 9.3 (bottom) shows a phased array probe is used on the shoulder of a chain link to see if there is any flaw in the weld zone. After final inspection, the chain is ready for delivery.

## 9.3 Wire rope

Wire ropes have a lighter weight and a higher elasticity than chain of the same breaking load. Designers use them in the makeup of mooring lines when all-chain designs become too heavy in deeper water. Common wire ropes used in offshore mooring lines are six-strand, eight-strand, and spiral strand, as shown in Figs. 9.4 and 9.5. The six-strand and eight-strand ropes are easier to handle due to their flexibility to bend on sheaves, and therefore are used more in temporary moorings. Spiral strand is torque neutral and can have a protective polyurethane sheath, and therefore it is more suitable for permanent moorings.

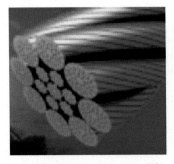

**FIGURE 9.4** Eight-strand wire rope (left) and sheathed spiral strand wire rope (right). *Courtesy: Bridon-Bekaert.*

**High torque and rotation**

6 Strand IWRC        8 Strand IWRC

**Low/Non rotating**

Spiral strand        Sheathed spiral strand

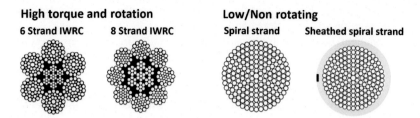

**FIGURE 9.5** Typical wire rope constructions.

### 9.3.1 Six-strand versus spiral strand

Wire ropes can be of various constructions. Some of them consist of several strands, usually six or eight, laid helically around a center core to form the rope, as shown in Fig. 9.4 and Fig. 9.6. Each strand is made up of several wires laid helically in one or more layers. The number of strands and wires in each strand (i.e., $6 \times 36$, $6 \times 42$, $6 \times 54$), core design, and lay of strands are governed by required strength and bending fatigue considerations for the rope. The center core has three types of designs: fiber core, wire strand core, or independent wire rope core (IWRC). For offshore mooring applications, IWRC ropes are used due to their durability.

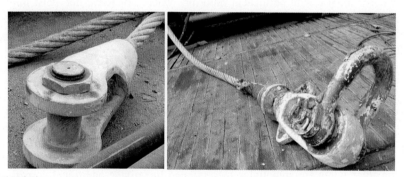

**FIGURE 9.6** Open-end socket on a wire rope (left) and close-end socket on a wire rope (right).

Six-strand ropes are most commonly used in temporary moorings, because they can be easily handled with their flexibility to bend. However, this type of construction generates torque as tension increases. Designers need to be careful when using these six-strand ropes in conjunction with polyester ropes. For such a conjunction, the six-strand wire rope under tension would want to unwind itself (untwist its six strands) using the entire polyester rope next to it as a pseudo swivel. The possible result is that the wire rope may experience a torsional fatigue after many cycles of tension.

To stop the torque from imparting line twist, torque-matched polyester ropes may need to be special-ordered and used.

Permanent moorings use mostly spiral strand wire ropes, as shown in Fig. 9.5. Spiral strand type uses one single strand containing a large number of wires. It is normally sheathed for corrosion protection. Sheathed spiral-strand wire consists of a steel wire core encased in a medium-density polyethylene jacket or sheath that seals the wire and protects it from corrosion. This allows the rope to have a long service life. However, the outer sheathing is susceptible to damage from handling during installation. A cut, abrasion, or reduction in thickness that exposes the steel wire to seawater will lead to corrosion and a shorter service life. Damage that does not penetrate completely through the jacket may propagate over time, as the mooring line flexes from cyclic loading, and eventually expose the steel wire to seawater. Sheathed wire ropes need to be handled with care. It is also possible to repair the sheathing and repair kits are typically available during installation.

For permanent moorings, another similar type of construction is multistrand. It does not have the sheath that is normally fitted on spiral strand type, and therefore is not recommended for use in a facility with a long design life. Both types have constructions that do not generate significant torque with tension changes. These spin-resistant (i.e., torque neutral) constructions are attractive for use with permanent moorings where imparting line twist might cause issues such as torsional fatigue or chain hockling (knotting). Both types of constructions use layers of wires (or bundles of wires) wound in opposing directions to obtain the spin resistance characteristics.

## 9.3.2 Corrosion protection

For corrosion resistance in permanent moorings, typically a high-density polyethylene or polyurethane jacketing is employed. The jacket (or sheath) is normally yellow as shown in Figs. 9.4 and 9.5 for better underwater visibility. Also, all wires can be galvanized, with or without the use of sheathing. Zinc filler wires are sometimes incorporated to provide additional corrosion protection as well. A blocking compound, that is essentially grease, is used as a lubricant and to block the inside spaces between the wires to minimize the spread of corrosion due to ingress of salt water.

Based on corrosion resistance, typical life expectancy of different types of wire ropes in permanent systems is recommended below by API Standards RP-2SK [7].

| | |
|---|---|
| Galvanized 6- or 8-strand | 6−8 years |
| Galvanized unjacketed spiral strand | 10−12 years |
| Galvanized unjacketed spiral strand with zinc filler wires | 15−17 years |
| Galvanized jacketed spiral strand | 20−25 years |
| Galvanized jacketed spiral strand with zinc filler wires | 30−35 years |

A recent study [8] found that the corrosion endurance of steel wire rope mooring lines is largely driven by the longevity of the galvanizing and the blocking compound, which forestall the direct corrosion loss of metallic area of the relatively small steel wires. Zinc filler (anode) wires may not serve much of a role in protecting exposed steel, as long as the galvanizing and the blocking compound are functional.

### 9.3.3   Termination with sockets

Wire rope is terminated with a socket for connection to the other components in the mooring line. The socket can be either an open socket or closed socket design. Fig. 9.6 displays an open-end socket on a sheathed spiral strand wire rope; while the right one shows a close-end socket on an eight-strand wire rope. The closed-end socket requires a shackle for connecting to the next line segment, likely a chain, while the open-end socket can be designed to connect directly to the chain link.

To connect the wire rope socket to the end of the wire rope, a poured socket is used to make a high strength, permanent termination. The poured socket is created by inserting the wire rope into the narrow end of a conical cavity (socket). The individual wires are then splayed out inside the cone, and the cone is then filled either with molten zinc, or more commonly, an epoxy resin compound. For permanent moorings, the sockets are typically provided with bend stiffeners (bend restricting boots) joined to the socket in a manner to seal out the ingress of water and limit free bending fatigue. Zinc anodes are often attached to protect the socket from corrosion. Typically, the socket is electrically isolated from the rope.

When a wire rope is put under load, its length increases by a small amount, which is the sum of the constructional extension and the elastic stretch of wire material. Constructional extension is caused by the wires adjusting themselves to their proper position in the strands to fill any gaps between the strands. As the amount of elongation is small, it is usually neglected in a mooring analysis.

### 9.4   Polyester rope

Polyester rope has become the choice of line types for deepwater permanent mooring applications due to its light weight and high elasticity. The elasticity of polyester ropes has allowed the use of taut systems in deep and ultradeep water without the need for catenary compliance to limit dynamic tensions, mostly excited by vessel motions due to waves. It has been extensively used in permanent moorings in deep water. Mooring analysis studies showed that polyester rope has desirable elasticity and stretch characteristics for mooring systems in the 1000−3000 m (i.e., roughly 3000−10,000 ft) water depth range. In deeper waters, a polyester mooring system can maintain a smaller vessel offset than a steel chain—wire—chain system. In some cases, the use of polyester rope could result in the use of fewer mooring lines for the same mooring performance when compared to a chain—wire—chain system for the same facility.

Additional benefits of using polyester moorings include reduction in hull structural costs due to smaller vertical loads and reduction in the extreme line dynamic tension due to lower stiffness. In summary, polyester rope offers these four advantages: reduced vessel offset, smaller mooring footprint, improved vessel payload capacity, and excellent fatigue properties.

While polyester ropes are widely used in deepwater permanent mooring applications, they have also seen increased use in preset moorings and in extending the water depth range of MODUs [9]. Fig. 9.7 shows a torque-matched polyester rope on a reel ready to be delivered to a client for mooring a tender assisted drilling unit.

**FIGURE 9.7**   Polyester rope on a reel ready to be delivered.

## 9.4.1    First use of polyester mooring in deepwater

Petrobras, the Brazilian national oil company, pioneered the use of polyester ropes in deepwater moorings in 1995 [10]. A 300-m section of polyester rope was successfully installed in one mooring leg of a platform. It was the first application of large polyester rope in deepwater moorings. After 1 year of service, the rope was removed, examined, and tested [10]. It was found that the residual strength remained the same as new. Since then, Petrobras has installed several permanent polyester mooring systems in the following years. The first deepwater mooring system to employ polyester in all mooring legs was installed in 1997. The use of polyester eliminates the need for steel wire ropes in those applications.

The successful uses of polyester ropes in deep water may be attributed to several research efforts in the 1980s and early 1990s. One of the most notable efforts is the work done by Del Vecchio [11,12]. In his study, polyester ropes were analyzed in a comprehensive way. It included assessment of load−elongation behavior, potential failure modes, and design methodologies. The study was partially based on model ropes. Confirmation of load−elongation behavior in full scale and fatigue performance data were obtained in the following years. His effort eventually led to the first installation of one polyester rope in 1995.

## 9.4.2    Rope constructions

Polyester ropes are typically constructed of several smaller subropes laid in a parallel construction. They are wrapped with soil filter to block soil ingress and a jacket to protect from abrasion. An example is shown in Fig. 9.8. The bundle of subropes forms a core to withstand tensile loads, with a jacket that is not load-bearing.

Construction types suitable for the subropes can be (1) parallel; (2) braided; or (3) laid. They can be used in different combinations depending on the requirements. In general, parallel strand keeps load-bearing yarns more aligned with the rope axis. The interaction between the fibers or strands will be low, and therefore it tends to get a higher strength efficiency. However, a small amount of twist may be desirable to give structure to the yarns and strands and to enhance load sharing among the components that make up the rope. For braided ropes, half the strands have a clockwise orientation and the other half have a counterclockwise orientation. The interaction between the strands is a point contact. This gives a rotation-free rope with excellent handling characteristics. However, the point contact between the strands influences fatigue and strength. A laid subrope typically will have all strands oriented in one direction introducing line contact between the strands. This provides excellent fatigue performance, both in tension and bending.

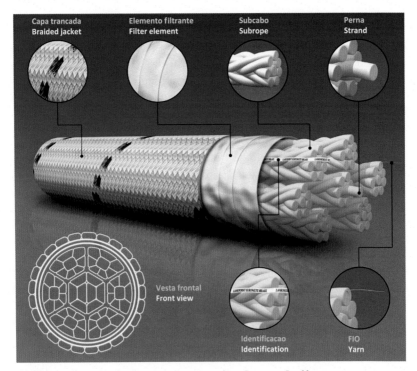

**FIGURE 9.8** Example of polyester rope construction. *Courtesy: Lankhorst.*

However, these ropes will rotate under load. Similar to steel wire ropes, this rotation problem can be overcome in a laid rope by using nonrotating designs.

Polyester ropes are typically constructed to be torque neutral. A torque-matched rope is sometimes used, when it is connected to a six-strand wire rope which is not torque-free.

To enhance the polyester rope's performance, rope manufacturers typically apply a nonwater-soluble marine finish coating to fibers. The purpose of the marine finish includes providing lubrication to assist bedding-in of the rope during initial tensioning, and increasing the rope's service life by reducing yarn-on-yarn abrasion.

A braided jacket is used to guard the rope core against damage from external abrasion that can occur while in service and during transportation, installation, and recovery. The jacket has little to no tensile load bearing capability. Risk of damage to the load bearing cores due to external abrasion may be reduced by the appropriate selection of a jacket design. This outer jacket also holds the subropes together to maintain the geometrical shape. A clearly visible color marking (e.g., painted straight line) on the jacket is typically provided to allow twist monitoring of the rope. A soil ingress protection layer, that is, soil filter, is often placed between the jacket and load-bearing fibers to give additional protection against soil ingress and marine growth.

### 9.4.3 Polyester stretch

Compared to steel wire rope and chain, polyester ropes have several special properties and requirements including the following:

- Construction stretch and creep.
- Nonlinear stiffness.
- Delicate handling procedures.
- Requirement to stay away from fairlead and seafloor.

Among those, construction stretch is one of the major drawbacks. Unlike steel wire rope and chain, polyester ropes exhibit axial load−elongation characteristics that are nonlinear, depending on loading type, and varying with time and loading history. The rope length after installation pretensioning will be longer than the manufactured length. Similarly, the length after the first significant loading (i.e., storm loading) will be longer than the installed length. The designer needs to be aware of these length changes, which are determined through load testing prior to manufacturing. In simpler words, they stretch significantly, and it is difficult to determine their lengths.

The underlying cause is the free-space between the filaments in the yarn. This free-space is reduced when the fibers start bedding in. This causes a reduction in diameter of the yarns, while the length of the filament does not change. Because of the helical structure, this results in an overall lengthening.

In addition, polyester fiber displays viscoelastic properties. As the rope is loaded beyond its previous maximum load, it undergoes a permanent increase in length due to material creep and construction stretch. The elongation of mooring lines results in larger mean offsets for the floating vessel. Vessel offsets are a major concern for the integrity of risers. Proper evaluation of offsets requires detailed information on the permanent elongation (bedding-in) and load−elongation properties of the ropes over a range of tensions. Permanent elongation, whether due to construction stretch or creep, can lead to the need to adjust the length of the mooring lines in the field.

### 9.5 Other synthetic ropes

Polyester [polyethylene terephthalate (PET)] is not the only fiber material that can be used to make mooring ropes. There are several fiber materials that can be considered for use in permanent or temporary moorings. These include nylon (polyamide), HMPE (high modulus polyethylene), aramid (aromatic polyamide), and others. As early as the 1970s, small nylon and aramid mooring lines were used successfully to moor instrumentation and navigation buoys in deep water. HMPE has been used recently on MODU moorings [13], and it is also used on permanent moorings in a few special cases. Currently, polyester is the most commonly used synthetic fiber for offshore mooring applications due to its low cost, light weight, low axial stiffness, and good fatigue properties. Other fiber materials can be more advantageous depending on situations and needs.

### 9.5.1  Nylon rope

Nylon is extremely elastic when compared with other types of materials used for moorings. For decades, nylon rope has been widely used as mooring lines for vessels alongside of piers, as towing hawsers, and as CALM buoy hawsers. It is used wherever high elasticity is a required property. These hawsers can be inspected frequently and replaced. Also in shallow water locations, a length of nylon rope can be inserted in the mooring line to absorb the energy from vessel dynamics. The Oil Companies International Marine Forum (OCIMF) conducted several joint industry projects (JIPs) in the late 1970s and early 1980s to study the properties of ropes used as conventional mooring lines. Different polyester, nylon, polypropylene, and polyethylene ropes were tested for dry and wet breaking strength, wet tension cyclic load fatigue, and external abrasion properties. The OCIMF single point mooring (SPM) Hawser Guidelines, which resulted from those JIPs, established procedures for specifying, prototype testing, manufacturing quality assurance, and inspection of large synthetic fiber rope [14,15]. These now serve as the basis for other fiber rope guidelines and test procedures.

### 9.5.2  High modulus polyethylene rope

HMPE has several properties superior to other fiber materials, such as excellent abrasion resistance, higher strength, and specific gravity less than seawater (i.e., it floats). HMPE ropes (used in Dyneema and Spectra brands) have seen some use in MODU moorings [16]. They are lighter, easier to handle, and have a smaller diameter than comparable polyester ropes of the same break strength. However, they may not be as cost-effective as polyester ropes in most mooring applications. Also, the conventional grades of HMPE may be more susceptible to creep and creep rupture. It may not be a concern for temporary mooring systems, but can be a concern for permanent deepwater moorings. Unlike the other synthetic fibers discussed here, the rate of creep of HMPE does not decrease logarithmically over time. HMPE Creep might continue at essentially a constant rate, requiring periodic retensioning of the mooring legs. As a result, the rate of creep may increase and potentially cause relatively sudden rope failure. Also, the rate of creep increases at higher temperatures.

A recent development has introduced new grades of HMPE [16]. These advanced grades of HMPE yarn are processed to greatly reduce creep. However, the rate of creep can still increase with temperature. Thus the rate of creep of these grades of HMPE may be acceptable at the relatively low temperature experienced when they are submerged in seawater.

HMPE ropes have been used successfully in temporary deepwater moorings on semisubmersible MODUs and they have also been used successfully as safety lines for floating production storage and offloadings. On one occasion, HMPE ropes were used to back up some highly corroded steel wire ropes. In another application, they were used as the mud lines (pig tails) of safety anchors that were retrofitted to supplement some compromised suction piles. With its superior abrasion resistance, HMPE has also been used to make cut-resistance jackets for polyester mooring ropes deployed in areas of trawl fishing. The special jacket was tested in a lab to simulate the clashing and sawing by a long steel trawl wire rope. As a result of the testing, polyester ropes with Dyneema jackets may see more deployments in the future.

In a recent application of the permanent mooring of a mid-water riser arch buoy, the use of HMPE tethers demonstrated several performance advantages over chain and steel wire tethers. Being lighter, the HMPE tether ropes reduced the size of the buoy and allowed for an easier installation. They also eliminated the issues with corrosion and fatigue. Fig. 9.9 shows the arrangement.

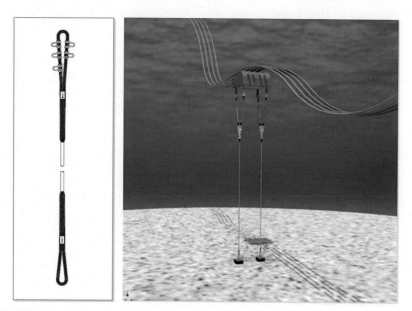

**FIGURE 9.9** Mid-water riser arch buoy permanently moored by HMPE ropes. *HMPE,* High modulus polyethylene. *Courtesy: Lankhorst.*

### 9.5.3 Aramid rope

Aramid (used in Kevlar and Twaron brands) rope has strength and stiffness comparable to steel wire rope. It is only occasionally used for offshore moorings, because it has a failure mode of axial compression fatigue, which can cause the rope to fail if fibers are subjected to compression. The first attempt to use large aramid fiber rope in a deepwater mooring system was in 1983 [9]. These aramid ropes were break tested to verify their strength. The aramid mooring lines were preset several months before a barge arrived in a manner that allowed them to become slack and rotate. This action caused axial compression fatigue in and near the bottom splices, which reduced the rope strength. When the barge was finally moored and the lines were tensioned to set the anchors, several of the aramid ropes failed. Prior to that incident, there was little understanding of axial compression fatigue. Note however that the axial compression can be avoided or minimized by good rope design, by maintaining a tension in the rope that does not allow fibers to see compression, and by using good termination techniques.

### 9.5.4 Considerations for moorings in ultradeep waters

Conventional polyester fiber ropes have provided an appropriate stiffness for existing deepwater moorings. As mooring line lengths increase in much deeper water, stiffer ropes may be desired. Also, polyester rope has bigger size compared with HMPE and Aramid ropes. Installation of a large rope in ultradeep waters can be a significant challenge. In this case, perhaps one should look into the high strength fibers, for example, HMPE or Aramid.

Table 9.1 compares the approximate stiffness of ropes made of polyester fibers and of the high modulus fibers, that is, aramid and HMPE. When loaded in a static manner, ropes made of these materials will be approximately three or four times stiffer than conventional polyester ropes.

**TABLE 9.1 Typical values of rope stiffness in catalog break strength.**

| Rope type | Static stiffness | Low frequency dynamic | Wave frequency dynamic |
| --- | --- | --- | --- |
| Polyester | ~10 (5–35) | 15–40 | 15–40 |
| Aramid | 33 | 33 – 60 | 60 |
| HMPE | 35 | 35 – 70 | 70 |

*HMPE*, High modulus polyethylene.

Storage and handling of the longer lengths of rope required in mooring systems for ultradeep waters is another issue which may lead to consideration of the high modulus materials. For the same strength, the diameters of ropes made of high modulus fibers will be much smaller, at roughly two-thirds the diameter of polyester. This would permit much more rope length to be stored on the same drum, which is a consideration for shipping and installation. Not only do high modulus fibers have smaller size for shipping and handling, but they are also lighter than polyester ropes. Aramid rope is about 40% and HMPE rope is about 30% of the weight of an equivalent polyester rope. Their smaller size and lighter weight reduce the lifting equipment needed to handle the same length of rope.

There has been extensive research and development done on polyester for deepwater mooring applications. By contrast, there have been fewer and less extensive studies on the high modulus fibers and ropes than for polyester. These high modulus ropes, such as HMPE and aramid, can be very promising, and are yet to be tried and proven in deepwater mooring applications.

## 9.6 Connectors

For years, different types of connectors have been used for connecting adjacent mooring line segments, such as shackles, Kenter links, pear links, C-links, and others. Many of them have stress concentration points in their geometries, and therefore are allowed to be used only in temporary mooring systems due to the limited fatigue lives.

For permanent mooring systems, D-shackles and H-links are two types commonly used to make connections between mooring line segments. Because inspection and replacement of connecters in a permanent mooring are difficult, their designs need to be robust with adequate fracture toughness, fatigue life, and corrosion protection. Manufacturing of connecting hardware should be subject to an appropriate level of quality assurance corresponding to the same quality as offshore mooring chain.

### 9.6.1 Connectors for permanent moorings

D-shackles—The shackle is a connector that is very common in the offshore industry. It consists of a bow, which is closed by a pin. Many different types of shackles are available, depending on the application. The shackle can be used in both temporary and permanent moorings. Fig. 9.10 shows a D-shackle being forged.

H-Link—Named after their shape, H-links serve to connect two lengths of mooring line whether it is chain to chain, chain to wire rope, chain to polyester rope, or polyester rope to polyester rope. This type of connector was introduced to avoid time-consuming handling that is associated with

**FIGURE 9.10**  A D-shackle is being forged. *Courtesy: Vicinay and Asian Star.*

D-shackles and allows for mooring line segments with different sizes to be easily connected to one another. Fig. 9.11 shows an H-link connecting a chain segment to a polyester rope segment.

**FIGURE 9.11**  H-link connecting polyester rope and chain. *Courtesy: Vicinay.*

Subsea mooring connector (SMC) tools—Advanced subsea connectors were developed to allow easy connection and disconnection of two mooring line segments underwater. Their most common usage is to allow the subsea connection between an anchor chain attached to a pre-installed pile and the mooring line being deployed from the installation vessel. A variety of these tools exist, mostly custom built. Some have a male part and a female receptacle, which are installed onto two different line segments to be connected. Fig. 9.12 shows a photo of the male part of a subsea connector being deployed in the field. This particular design uses a ball-and-taper mechanism to lock the connection. The underwater operation of connecting or disconnecting the male and the female parts are performed by a remotely operated vehicle (ROV). An alternative design called ROV-operated SMC uses a conventional H-link mechanism with an improvement that enables an ROV to handle and close a pin.

**FIGURE 9.12**   Subsea connector on a temporary mud mat (left); male part ready to be deployed (right). *Courtesy: First Subsea.*

## 9.6.2   Connectors for temporary moorings

Kenter link—The Kenter connecting link is most commonly used for the connection of two pieces of mooring chain, where the terminations of the two chain pieces have the same dimensions. It has the same outside length as a chain link of the same diameter (Fig. 9.13). However, the width is larger than the common link and the diameter at the midsection is also larger than the common link. Both may be potential issues with installation and handling if used on a chain gypsy or coming through a chain stopper. Generally, Kenter links are not used in permanent mooring systems due to a shorter fatigue life than the chain.

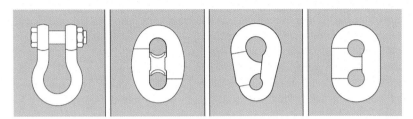

**FIGURE 9.13**  Shackle, Kenter link, pear link, and C-link. *Courtesy: Vryhof.*

Pear link (and Trident link)—The pear-shaped connecting link is similar to the Kenter link and C-link, except that it is used for the connection of two pieces of mooring line with terminations that have different dimensions. Like the Kenter link and C-link, the pear-shaped connecting links are not used in permanent mooring systems (Fig. 9.13).

There is a product called a Trident link that combines the features of Kenter and pear links. Trident link is mainly used to connect a chain directly to an anchor shackle, thus reducing the amount of required connections. It is very similar to a pear link with the only notable differences being the method of assembly. This type of shackle can also be used in a variety of other applications where the cross over from one size to another is required.

C-link—Like the Kenter link, C-link is used for the connection of two pieces of mooring line with terminations that have the same dimensions. The major difference between the Kenter type and the C type is the way that the connector is opened and closed. This connector is generally not used in permanent moorings (Fig. 9.14).

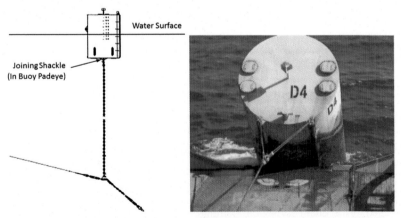

**FIGURE 9.14**  A surface buoy of a steel cylinder shape (left); a buoy is retrieved for maintenance (right).

Swivel—A swivel can be used in a mooring system (usually for temporary systems only) to relieve the twist and torque that builds up in the mooring line. The swivel is often placed a few links from the anchor point, although it can also be placed between a section of chain and a section of wire rope. There are many different types of swivels available. However, a downside to many common swivels is that while under high loads, they lock up due to the high friction inside the turning mechanism. Some newly designed swivels have a special bearing inside, and are capable of swiveling under higher loads.

## 9.7    Buoy

Surface or subsurface buoys can be connected to a mooring line. One of the purposes for adding a buoy is to increase vertical clearance between the mooring line and any subsea equipment such as pipelines or even mooring lines from another floater in proximity. Other benefits of buoys include improved mooring performance (reduced vessel offset) and reduced weight of mooring lines that must be supported by the vessel hull. However, integrity issues can arise because of the extra connections. Those issues can easily outweigh the benefits, so designers normally try to avoid them in permanent moorings unless it is necessary.

Buoys can be placed in line with the mooring (with a strength member passing through the buoy) or attached separately to the mooring line through a tri-plate. When using the inline buoy approach, care must be taken to allow for rotation in the end connections. Also, since it is in the load path of the mooring line, its strength needs to be designed accordingly. The other approach is to attach a buoy separately like a pendant. Fig. 9.14 shows an example of a pendant approach where the buoy is designed to stay on water surface. A tri-plate is used in the mooring line for connecting the pendant chain.

Buoys used with permanent moorings are typically constructed from syntactic foam, steel, or a combination of synthetic material surrounding a steel structure. Syntactic foam is a material made of glass/carbon spheres encased in high-density foam. Its depth rating is mainly determined by the pressure rating of the glass/carbon spheres. It has been widely used to provide buoyancy for deepwater drilling and production risers as well as mooring operations.

Steel buoys may provide a cost-competitive solution, especially when their sizes are relatively large (see the example in Fig. 9.14). The buoys can be built in either a cylindrical or a spherical shape. The former can use ring stiffeners in its body, and the latter may use unstiffened dished ends welded together. Subsurface spring buoys of steel construction need to be designed for external pressure according to recognized pressure containment standards. They are designed to have adequate strength for maximum operating depth.

Using buoys can have adverse effects. Buoys require additional connecting hardware which increase installation complexity. Also, it can potentially increase design loads on the mooring lines due to dynamic response of the buoy in heavy seas, especially in the case of surface buoys. The induced loads can be very dynamic because a buoy tends to have significant motion in waves due to its small size. Surface buoys often break away when their connecting hardware eventually fails in fatigue or wear due to this continued motion. Fig. 9.14 shows a buoy getting retrieved for maintenance. In that particular case, the padeye and shackle underneath were severely worn to the point that a repair was required.

## 9.8 Clump weight

Clump weights are sometimes fitted on ground chains to improve mooring performance, particularly in reducing vessel offset. The additional weight from these cast steels can increase the restoring force of the mooring system, as the vessel would have to lift those weights before it can offset further. If used, they are typically added to a short segment of ground chain near the touch down point to increase the restoring force of a mooring leg. A study [17] has demonstrated that, compared with a conventional mooring system with uniform lines, a mooring system with optimum clump weight design can have improved performance characteristics. The study investigated design parameters of clump weights and their positive effects on mooring system performance. Note that clump weights can have integrity issues if they are not carefully designed. They have a tendency to break loose or eventually fall apart after years of beating up and down in the touch down zone. Fig. 9.15 shows clump weights of a mono-cast design which

**FIGURE 9.15** Clump weights of a mono-cast (single-shell) design fitted on chain.

may provide a better durability than the conventional design with two half shells bolted together.

In some alternative designs, a segment of a much-larger chain is intentionally used to serve the purpose of clump weights. Also, parallel chains ended with large tri-places can be another alternative way to avoid the need of clump weights.

## 9.9   Questions

1. Studlink chain is heavier than studless chain. Why do people still use studlink chain rather than studless chain for mobile moorings and ship anchoring?
2. Six-strand wire ropes do not have sheaths (jackets) that provide corrosion protection. Why are they still so commonly used for temporary (mobile) moorings?
3. Name one advantage and one drawback of polyester mooring ropes.
4. What is the reason that Kenter links are not used in permanent mooring systems?
5. Name at least one reason why a buoy is used. Also, explain why buoys can bring problems.

## References

[1]   ISO 20438, Ships and Marine Technology—Offshore Mooring Chains, ISO International Standards, 2017.

[2]   DNV CN 2.6, Certification of offshore mooring chain, in: Det Norske Veritas Certification Note No. 2.6, July 1985.

[3]   DNV CN 2.6, Certification of offshore mooring chain, in: Det Norske Veritas, Certification Note No. 2.6, August 1995.

[4]   API Spec 2F, Specification for Mooring Chain, first ed. issued 1974, third ed., January 1981, American Petroleum Institute, 1981.

[5]   Act, The anchors and chain cables rules 1970, in: Statutory Instrument 1970 No. 1453, Act of English Parliament, 1st October 1970.

[6]   A. Potts, G. Farrow, A. Kilner, Investigations into break strength of offshore mooring chains, in: OTC-27678, Offshore Technology Conference, May 2017.

[7]   API RP-2SK, Design and analysis of stationkeeping systems for floating structures, API Recommended Practice 2SK, third ed., American Petroleum Institute, 2005.

[8]   J. Rosen, A. Potts, E. Fontaine, K. Ma, R. Chaplin, W. Storesund, SCORCH JIP—feedback from field recovered mooring wire ropes, in: OTC 25282, OTC Conference, May 2014.

[9]   J.F. Flory, S.J. Banfield, C. Berryman, Polyester mooring lines on platforms and MODUs in deep water, in: OTC 18768, Offshore Technology Conference, 2007.

[10]   R. Rossi, C. Del Vecchio, R. Goncalves, Moorings with polyester ropes in petrobras: experience and the evolution of life cycle management, in: OTC 20845, Offshore Technology Conference, May 2010.

[11]   C. Del Vecchio, Light Weight Materials for Deep Water Moorings (Ph.D. dissertation), University of Reading, 1992.

[12] C.R. Chaplin, C. Del Vecchio, Appraisal of lightweight moorings for deep water, in: OTC 6965, 1992.

[13] S. Leite, J. Boesten, HMPE mooring lines for deepwater MODUs, in: OTC-22486, OTC Brasil, 2011.

[14] OCIMF (Oil Companies International Marine Forum), Guidelines for the Purchasing and Testing of SPM Hawsers, first ed., 2000.

[15] OCIMF (Oil Companies International Marine Forum), Recommendations for Equipment Employed in the Bow Mooring of Conventional Tankers at Single Point Moorings, fourth ed., 2007.

[16] M. Vlasblom, J. Boesten, S. Leite, P. Davies, Development of HMPE fiber for permanent deepwater offshore mooring, in: OTC 23333, Offshore Technology Conference, May 2012.

[17] Y. Luo, Optimum design of clump weights for offshore mooring systems, in: Proceedings of ISOPE, International Society of Offshore and Polar Engineers, 1992.

# Chapter 10

# Hardware—on-vessel equipment

## Chapter Outline

10.1 **Tensioning systems** 199
    10.1.1 Fairlead and stopper 201
    10.1.2 Hydraulic or electric
        power unit 201
    10.1.3 Chain locker 202
10.2 **Chain jack** 203
10.3 **Chain windlass** 204
    10.3.1 Movable windlass (or
        chain jack) 204

10.4 **Wire winch** 206
    10.4.1 Drum winch 206
    10.4.2 Traction winch 207
    10.4.3 Linear winch 209
10.5 **In-line tensioner** 209
10.6 **Summary** 212
10.7 **Questions** 212
**References** 212

This chapter introduces on-vessel equipment which is mainly the tensioning system on the deck of the hull structure and the fairleads somewhere underneath it. The selection of a suitable tensioning system is one part of the design process in mooring system engineering. To provide an understanding of these equipment, types of mooring tensioning systems are reviewed, including chain jack, chain windlass, and wire winches. An in-line tensioner is also introduced as it serves the function of tensioning like the other types, even though it is physically off-vessel rather than on-vessel.

## 10.1 Tensioning systems

Mooring tensioning systems have evolved from rotary windlasses on ships into multiple options nowadays. These options include various types of winching equipment that could be linear or rotary, electrically or hydraulically driven, and fixed or movable. An alternative option which has emerged recently is to use an in-line tensioner and remove the winching equipment from the deck completely. Another option is to use a portable winch which can be removed after the mooring installation; the option is popular for some cost-sensitive industries such as offshore floating wind farm.

    The primary function of a tensioning system is to provide necessary pull-in force to perform the mooring system hook-up within acceptable time limits during installation. Other functions include allowing pull-in and

Mooring System Engineering for Offshore Structures. DOI: https://doi.org/10.1016/B978-0-12-818551-3.00010-7

pay-out when the vessel needs to be repositioned or when the first chain link off the stopper needs to be shifted (i.e., refreshed, rotated).

There are several types of tensioning systems. They can be generally classified as *chain jack*, *chain windlass*, or *wire winch*. These three types are further introduced in the following sections. A complete mooring tensioning system includes the tensioning equipment, fairleads (Fig. 10.1), chain stoppers, and power unit, which can be either hydraulically or electrically driven.

**FIGURE 10.1** Windlass (top) and fairlead (bottom) for mooring chain. *Courtesy Rolls-Royce.*

### 10.1.1 Fairlead and stopper

Mooring lines are subjected to high wear rates and stresses at the fairlead and stopper arrangements. The long-term service of a mooring system requires that fairlead and stopper arrangements be designed to minimize wear and fatigue. For example, fairleads should provide sufficient sheave-to-rope diameter ratio (i.e., $D/d$ ratio) to minimize tension-bending fatigue on wire ropes. Typically, sheaves for wire rope have $D/d$ ratios of 16−25 for mobile moorings, and 40−60 for permanent moorings. For chain, seven to nine pocket wildcat sheaves are typically used. Mooring chain is often stopped off at the vessel's hull in order to take direct mooring loads off the winch. Chain stoppers (Fig. 10.2) and wire rope grips are designed such that the stress concentrations and wear within the chain or wire rope are kept at acceptable levels.

**FIGURE 10.2**  Chain stopper closed and opened. *Courtesy NOV-BLM.*

Fig. 10.3 shows a fairlead chain stopper of an advanced design. The fairlead has a long arm that can pitch and yaw (swing horizontally) freely around the two axes at its base. This double-articulation design allows the fairlead system to move like a universal joint. The chain stopper is placed at the outer end of the arm. Such an arrangement encourages the rotations to take place in the double axes (articulations); therefore, the interlink wear and chain out-of-plane bending may be reduced.

### 10.1.2 Hydraulic or electric power unit

Mooring tensioning systems can be either hydraulic or electrically driven. Chain jacks and linear winches are typically hydraulically driven. Although an electrically driven chain jack is technically feasible, it will require an extra mechanical transmission system and may be costly and space-consuming. Rotary winches can be either electrically or hydraulically driven.

For a hydraulic drive, a hydraulic power unit (HPU) typically comprises several pump groups. This arrangement builds in redundancy that, in the

**FIGURE 10.3**    Fairlead of a double-articulation design. *Courtesy MacGregor.*

event of failure of one pump group, the winch system can still be operated at the same pulling capacity, albeit at reduced speeds. Compared to electrical drive, hydraulic drive typically has advantages in a simplified gearbox and higher degree of redundancy due to multiple hydraulic motor-pump units. It also provides extra hydraulic braking. The disadvantage of hydraulic drive is that it requires more maintenance, together with greater installation cost, and the risk of hydraulic fluid leakage.

Electric drive is heavily used in drilling and accommodation platforms, which need frequent mooring line pull-in and pay-out when the platform relocates. For an electric solution, there will be no need for a hydraulic system or components to operate the winches. An electric drive can be combined with a rotary windlass. Compared to a hydraulic drive, the electric drive has advantages in less maintenance, lower installation cost (electrical cables vs hydraulic piping), less noise, and slightly lower weight. Electric drive also allows for easy, continuous operation even in the pay-out mode, and it is 100% reversible with reversed load coming from the chain locker. An electric solution is also more efficient in terms of energy consumption compared with hydraulic units.

### 10.1.3 Chain locker

Offshore platforms or vessels may have chain lockers for storing chain. A chain locker can be simply a compartment located under the chain jack or windlass where the platform chains are stowed. Chain lockers can also be located on the deck level if the platform has limited space inside of the hull, although such an arrangement is not favorable for the center of gravity of the floating structures. Chain lockers can be made watertight on their openings to ensure that the watertight integrity is maintained and to avoid sudden flooding of the chain lockers. The openings are equipped with closing devices.

## 10.2   Chain jack

A chain jack is a device which reciprocates linearly to haul-in and tension chain. Fig. 10.4 shows examples of typical chain jacks. Usually powered by one or more hydraulic cylinders, a chain jack engages the chain, pulls in a short amount of the chain, engages a stop, retracts, and repeats the process. Although a chain jack can be a powerful means for tensioning chain, it is very slow and is recommended for applications not requiring frequent heaving-in or paying-out mooring lines such as permanent moorings.

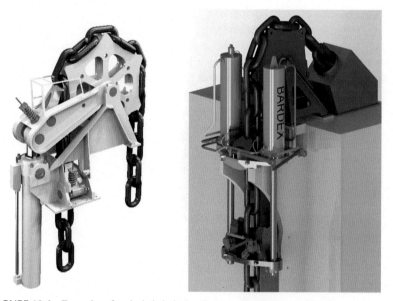

**FIGURE 10.4**   Examples of typical chain jacks. *Courtesy MacGregor and Bardex.*

Linear chain jacks come with turn-down sheaves to route the chain into the chain lockers. They often feed chain into chain lockers inside the hull to store excess chain during installation. Turn-down sheaves can be integrated with chain jacks, or separately sit behind chain jacks. A chain stopper secures the chain in position for the life of the service. It is typically self-closing (i.e., fail-safe). Control units usually include a local control console for automatic or manual control of the tensioning system. The control units synchronize the movements of components, including the chain jack pawls and cylinders, during normal heave-in and pay-out operations.

The pull-in capacity of a tensioning system will be the maximum pretension plus appropriate margin. Note that most mooring systems have pretensions set in the range of 10%−20% of the mooring line minimum breaking load (MBL). Stalling capacity needs to be at least higher than the expected highest proof load. Typically, the stall capacity is in the range of 1.2−2.0 times the highest design pretension or the required proof load during installation, whichever is higher. The Class Rule by Det Norske Veritas Germanischer Lloyd has a requirement of no less than 40% of mooring line MBL [1].

## 10.3 Chain windlass

Another method of handling and tensioning chain is through the use of a windlass, as shown in Fig. 10.5. The windlass consists of a slotted wildcat (i.e., a gypsy wheel) which is driven by a power source through a gear-reduction system. As the wildcat rotates, the chain meshes with the wildcat, is drawn over the top of the wildcat, and lowered into the chain locker. Once the chain is hauled in and tensioned, a chain stopper or brake is engaged to hold the chain. While a windlass may occupy larger footprint than a linear chain jack, it provides a fast and reliable method for handling and tensioning mooring chain in both pull-in and pay-out. Chain windlasses are widely used on MODUs.

**FIGURE 10.5** Two examples of typical fixed windlasses. *Courtesy MacGregor and Bardex.*

### 10.3.1 Movable windlass (or chain jack)

Having one windlass or chain jack dedicated for each mooring line is convenient but can be costly. For a floating facility with a larger number of mooring lines on each cluster, a movable windlass or chain jack can be attractive.

A movable tensioning system has a windlass or a chain jack sitting on a skidding beam (i.e., a rail track). The windlass or the chain jack can be repositioned to serve other mooring lines in the same cluster. Movable winches typically require additional material handling to prepare the windlass or chain jack for operation. There are mainly two methods to move the tensioner: (1) a handling frame on skidding beams, as illustrated in Figs. 10.6 and 10.7, or (2) a dedicated overhead crane. Both options can use chain hoists that are hydraulically, pneumatically, or electrically driven. Besides the two options, it is also possible to relocate the tensioner from one station to the next through platform pedestal cranes. However, the platform crane needs to have the required lifting capacity and boom range to reach all columns.

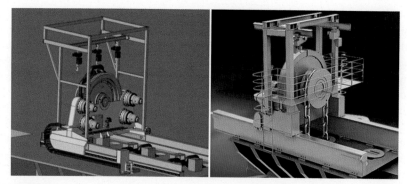

**FIGURE 10.6**  Movable windlass with handing frame on skidding beams. *Courtesy MacGregor and Rolls-Royce.*

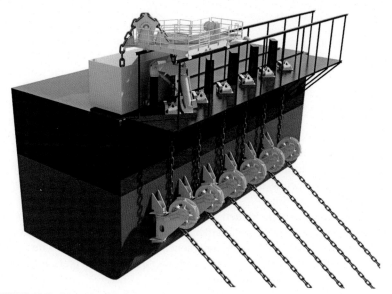

**FIGURE 10.7**  Movable chain jack. *Courtesy MacGregor.*

Movable winches (or chain jacks) have the advantage of reduced CAPEX (capital expenditure) due to fewer tensioners being required at each mooring cluster. However, it increases the complexity of a tensioning job and requires a larger team for the operations compared with fixed tensioners. TOTAL's Girassol FPSO (floating production storage and offloading) was the first floating facility to utilize a movable chain jack system back in 2001. Since then, multiple facilities have incorporated movable chain jacks into the design.

## 10.4 Wire winch

The preferred choice of the tensioning systems for ship-shaped FPSOs depends on whether the vessel is turret or spread moored. For a turret mooring system, all mooring lines are typically pulled-in and tensioned using a single *drum winch* located on or near the turret. The work wire on the winch goes through turn-down sheaves and pull-tubes to reach top chains held by chain stoppers located in the chain table at the bottom of the turret. For a spread moored FPSO, one drum winch can be placed at the bow with another at the stern; it is also popular to use moveable chain jacks with one chain jack serving a group of mooring lines. For moored mobile offshore drilling unit applications, tensioning systems are typically comprised of a wire winch and chain windlass for each mooring leg. The wire winch can either be a *drum winch* or *traction winch*, depending on capacity requirements and costs.

### 10.4.1 Drum winch

The conventional drum-type winch is a common method used for handling wire rope, as shown in Fig. 10.8. Operation of a drum-type winch is typically fast and smooth. A drum-type winch consists of a large drum on which the wire rope is wrapped. The tensioning capacity of the winch is a function of number of wraps on the drum. Pulling capacity is reduced with a full reel of wire; it increases when less wire is on the drum. The base of the drum is often fitted with special grooves sized specifically to the size of wire rope being handled. The groves control the positioning of the bottom layer of wire rope on the drum. For subsequent layers of wire rope, an external guidance mechanism such as a level-wind is often used to control positioning of the wire rope on the drum.

A drum winch may be limited for use in deepwater moorings or moorings with high tensions. As the requirement for line sizes and lengths increases, the size of the winch can become impractical. In addition, when wire rope is under tension at an outer layer on the drum, spreading of preceding layers can occur causing damage to the wire rope.

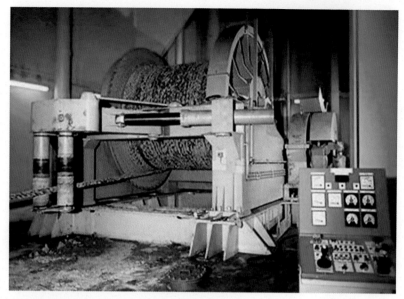

**FIGURE 10.8** Drum winch. *Courtesy MacGregor.*

## 10.4.2 Traction winch

The traction winch (as shown in Fig. 10.9) has been developed for high-tension mooring applications as well as for handling mooring systems of chain-and-wire combinations. It consists of two closely spaced parallel mounted powered drums, which are typically grooved. The wire rope makes several wraps (typically six to eight) around the parallel drum assembly. The friction between the wire rope and the drums provides the gripping force for the wire rope. The wire rope is coiled on a take-up reel, which is required to maintain a nominal level of tension in the wire rope (typically 3%−5% of working tension) to ensure the proper level of friction is maintained between the wire rope and the traction winch. This system has been favored for use in high-tension applications due to the compact size, capability to provide constant torque, and ability to handle very long wire rope without reduced pulling capacity. A traction winch has the same pulling capacity at any pay-out because the wire is spooled onto or off a storage reel and the traction heads provide the pulling capacity.

This type of tensioning system can also have an option for deployment of one additional mooring line from the same winch−windlass combo (winch package). For example, if there are eight traction winch packages on a facility for a 4 × 2 mooring system, this would allow the facility to deploy additional four lines to make it a 12-line mooring system, assuming the additional four fairleads are in place on the four columns of the hull. Fig. 10.10 shows an

**FIGURE 10.9**  Two traction winch packages on a semisubmersible rig.

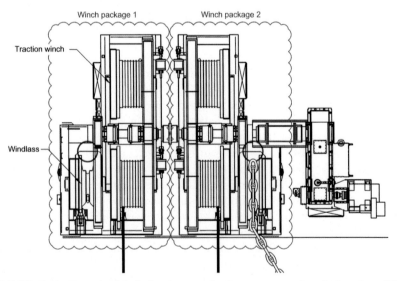

**FIGURE 10.10**  Example of deploying three mooring lines on two traction winch packages [2].

example of deploying three mooring lines from two winch packages, where each package consisting of one chain windlass and one traction winch. This means a four-column platform with an 8-point mooring system (4 × 2) can become a 12-point mooring system (4 × 3) without adding additional tensioning systems. The additional line is deployed using the chain windlass. While this unconventional setup may not be an optimized solution for all applications, it does provide a cost-effective means of increasing total mooring system capacity [2].

### 10.4.3 Linear winch

The linear winch is similar to a chain jack. Two sets of grippers, one stationary and one translating, are used to haul-in and tension the wire rope. A linear winch is available in a single-acting form, in which case the wire rope moves intermittently as the gripper is retracted to begin another stroke. It is also available in a continuous double-acting form, in which case two translating grippers are used alternately for continuous smooth motion of the wire rope. A linear winch is most applicable in a permanent mooring application, when high tension and large diameter wire rope is required. A take-up reel is necessary in this case to coil the wire rope after it passes through the linear winch. Note that experience has shown that wire ropes can easily get damaged by grippers, which may limit the use of this type of winch. Compared with chain jack, the linear winch has much more complicated deck arrangement and therefore is seldom used.

## 10.5 In-line tensioner

With an in-line tensioning system, tensioning of the mooring line is carried out using a boat on the sea surface that is equipped with a winch. The in-line tensioner itself is not an on-vessel device, but rather a permanent component in the mooring line, as shown in Fig. 10.11. By using a large wrap angle around the guide roller (wheel) in the in-line tensioner, the achieved tension in the mooring line will be higher than the pulling load applied by the winch on the boat. In simple analogy, the in-line tensioner serves as a single pulley with the anchor as the fixed point, and the boat can efficiently tension up the mooring line.

In a patent application, Dove and Treu [3] outlined the methodology for the use of an in-line mooring tensioner in 2001. The tensioner itself is defined as a chain wheel and a stopper held within a frame. The method of installing the mooring line was outlined in the patent; Fig. 10.11 illustrates the method schematically [3]. The idea was not adopted until the Stones FPSO project, in which an in-line tensioner was designed and deployed as a part of the Stones disconnectable turret mooring system [4,5].

Fig. 10.12 shows an example of in-line tensioner. Key components of the in-line tensioner include the chain stopper, which is located in the lower part

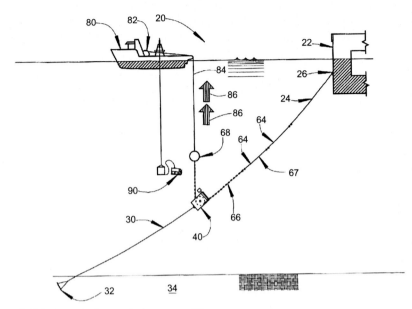

**FIGURE 10.11** In-line tensioning system [3].

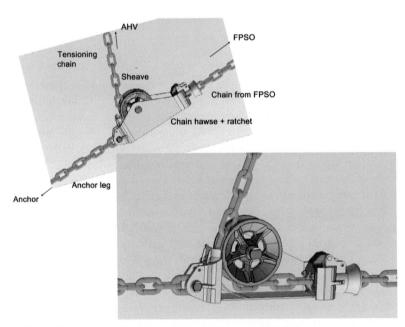

**FIGURE 10.12** Example of an in-line tensioner. *Courtesy SBM Offshore.*

of the tensioner, and the five-pocket chain wheel. The bi-axial fairlead joint is attached to the hull and provides both vertical and horizontal freedoms for the pitch and yaw movements of the mooring line.

Fig. 10.13 compares the differences between a conventional tensioning system and an in-line tensioning system. The most apparent advantage of the in-line tensioning system is the elimination of the large tensioning equipment on the platform. It translates into a reduced CAPEX and a reduced maintenance cost associated with the tensioning system onboard. Another significant advantage is that the mooring configuration with the in-line tensioning system does not have any platform chain, which means the splash zone corrosion on platform chains is completely eliminated. Note that splash zone corrosion has been one of the main integrity issues for permanent mooring systems (refer to Chapter 13: Mooring reliability, and Chapter 14: Integrity management, for more details on mooring reliability and integrity).

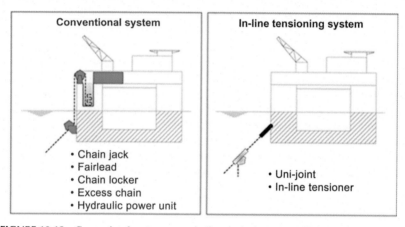

**FIGURE 10.13**    Conventional system versus in-line tensioning system [2].

The main disadvantage of the in-line tensioning method is the operating cost associated with making adjustments to any mooring line. During the hookup of the mooring system, there are appropriate vessels readily present in the field and the additional cost associated with the vessel time is limited. However, any adjustment to the mooring system later requires the mobilization of a vessel. Depending on the total number of adjustments needed over the field life, the associated operating costs may exceed the CAPEX savings from the elimination of the on-vessel equipment. For this reason, an in-line tensioning system may not be suitable for production platforms that need to relocate (reposition) periodically, for example, to shift the fatigue point in the riser at the touch down zone. Moreover, if the platform ever needs to relocate a longer distance, the remaining chain hanging under the in-line tensioner would be too short to allow it.

An in-line tensioner is placed somewhere in a mooring line, for example at the bottom end of the top chain. It can also be placed at the top of an anchor pile. This latter placement has been adopted by some mooring systems in shallow waters. With such a placement, the anchor padeye is moved from the conventional location to the top of the pile, which makes the pile less efficient in its holding capacity.

## 10.6 Summary

There are many choices when selecting a tensioning system. It is found that all alternatives (fixed or movable, electrically or hydraulically driven, and on-vessel or in-line tensioners) have their advantages and disadvantages. They need to be evaluated systematically to select the one that fits the purpose of the job and the requirement of the project.

A recent paper [2] compared different tensioning options for a production semisubmersible in a case study. It was found that the fixed hydraulic chain jack is the most popular choice for production semisubmersibles, based on the record in the past 24 years. The paper noted that the in-line tensioning system may be a promising alternative for certain projects because of its simplicity in configuration and the advantage of eliminating the splash zone corrosion problem.

It may be worth noting that on-vessel tensioning systems are responsible for some of the mooring accidents. For example, there was an accident on the Thunder Hawk semisubmersible in 2009, during which the failure of the tensioning system resulted in the dropping of one complete mooring line to the seafloor, including chain, rope, and connectors [6].

## 10.7 Questions

1. What are the three main types of conventional tensioning systems?
2. The power unit for a tensioning equipment can be either a hydraulic or electrical drive. Name at least one advantage for each.
3. What does the word "linear" mean when referring to linear chain jacks or linear winches?
4. Why is the tensioning capacity of a drum-type winch a function of number of wraps on the drum?
5. List at least one advantage and one disadvantage of the in-line tensioning system.

## References

[1] DNV GL, DNVGL-OS-E301. Offshore standard position mooring, July 2018.
[2] Y. Wu, T. Wang, K. Ma, C. Heyl, R. Garrity, J. Shelton, Mooring tensioning systems for offshore platforms: design, installation, and operating considerations, in: OTC 28720. Offshore Technology Conference, May 2018.

[3] P. Dove, J. Treu, Method of and apparatus for offshore mooring. US Patent No.: US6983714B2, Filing Date: 15 June 2001, Date of Patent: 10 January 2006.

[4] M. Macrae, Anchor line tensioning method. US Patent Publication No: US9340261B2, 17 May 2016.

[5] C. Webb, M. van Vugt, Offshore construction—installing the world's deepest FPSO development, in: OTC 27655. Offshore Technology Conference, Houston, TX, 2017.

[6] BSEE, United States Department of the Interior Minerals Management Service Gulf of Mexico Region Accident Investigation Report 090515, 22 July 2010.

# Chapter 11

# Installation

## Chapter Outline

11.1 **Site investigation** **215**
    11.1.1 Geophysical survey 216
    11.1.2 Geotechnical survey 216
11.2 **Installation of permanent mooring** **217**
    11.2.1 Phase I—installation of pile anchors 217
    11.2.2 Phase II—prelay of mooring lines on seabed 219
    11.2.3 Phase III—hook-up of mooring lines to floating production unit 222
11.3 **Deployment and retrieval of temporary mooring** **225**

11.3.1 Rig mooring system for mobile offshore drilling unit 226
11.3.2 Preset mooring system for mobile offshore drilling unit 228
11.4 **Installation vessel** **229**
    11.4.1 Anchor handling vessel 229
    11.4.2 Anchor handling vessel incident—capsizing of bourbon dolphin 230
11.5 **Questions** **231**
**References** **232**

This chapter introduces how permanent and mobile offshore drilling unit (MODU) moorings are installed in offshore fields. Before the engineering and installation phases of a project, geotechnical and geophysical surveys for the field are normally conducted. They provide essential information for determining the anchor locations and sizing the anchors. Some MODU operations, however, may simply use most available information without conducting any surveys. Going into the installation phase of the project, anchor handling vessels (AHVs) or construction/crane vessels are hired to do the job. The features of AHVs are reviewed as well.

## 11.1 Site investigation

Before a mooring installation is started, there should have been a site investigation that provides information about subsea terrain, topography (bathymetry), soil properties, etc. A site investigation campaign includes geophysical and geotechnical surveys that may cover subsea survey, positioning of facilities, soil sampling, and soil investigation [1,2]. It is one of the main activities for a field development. Information concerning the soil layering and properties need to be collected at and around the locations of the anchors and other subsea equipment. Survey vessels are sent to the site to conduct the surveys. Some of the vessels can cover both geophysical and geotechnical surveys. The vessels are normally equipped with an A-frame and heave-compensated offshore cranes that are capable of operating the required survey equipment. Site investigation is often performed early in a project before the engineering or installation phases.

### 11.1.1 Geophysical survey

The geophysical survey allows geohazards to be detected and also provides a regional geological overview. The purpose is to identify the potential man-made hazards (e.g., sunk ships, dropped or abandoned objects), natural hazards, and engineering constraints. It is also to assess any potential impact on biological communities and to determine the seabed and subbottom conditions. This survey should have a geological interpretation and be integrated with the possibly existing geotechnical data to assess restraints imposed on the design by geological features. Examples of potential geohazards are boulders, shallow faults, and debris flows [3] that can be a problem for anchors, mooring line routing, and laydown corridors for prelaid lines that will be installed at the particular location.

### 11.1.2 Geotechnical survey

The geotechnical survey allows for the collection of soil samples to determine soil properties for anchor design. A cone penetration test is conducted at the proposed anchor locations determined from review of the geophysical survey. In situ geotechnical tests, such as cone penetrometer tests and shear vane tests, are performed [1,2]. Based on the tests, the characteristics of the seabed soil around the field development area can be determined. The depth of the geotechnical borings should exceed the anchor's penetration. The number of these borings should be defined as a function of the soil variability. Normally, at least one boring is performed at each location of an anchor group (cluster). If there is a drastic change in one of the core samples, additional samples will be taken to determine the changes in condition. Soil properties, such as undrained shear strength, are determined for the design of pile anchors. For pile anchor design, a more extensive soil investigation is generally required, compared to drag embedment anchor design.

## 11.2 Installation of permanent mooring

A typical mooring installation procedure is introduced using a permanent floating production unit (FPU) as an example. The FPU has a hull shape of a semisubmersible. It is to be moored by a chain–polyester–chain system in deep water. A suction pile anchor is used as an example to show the installation procedure, as it is the most common anchor type for deepwater FPUs. From top to bottom, the mooring line profile consists of platform chain, polyester rope segments connected by H-Links, bottom chain, subsea connector, anchor chain, and suction pile.

A mooring installation can be divided into three phases, that is, anchor installation, mooring line prelay, and hook-up to the FPU. The three-phase approach has become one of the preferred methods in the industry, because it provides several advantages [4]. The breakdown of the three phases can make the overall project schedule more flexible. In the first phase, pile anchors are installed independently from the mooring lines. In the second phase, all mooring lines are completely installed and laid down on the seabed. In the third phase, the prelaid mooring lines are picked up from the seabed and connected to the hull. This approach allows the use of smaller vessels that are less expensive. It also reduces the complexity of the hook-up phase, as most of the connections between line segments were already made during the prelay phase [4]. However, expensive construction vessels are sometimes used because they will be required to install other equipment, such as risers, anyway.

### 11.2.1 Phase I—installation of pile anchors

Suction embedment anchors have arguably become the preferred anchor choice for permanent mooring applications in deep water. Essentially, they are large diameter piles with enclosed tops. They are lowered to self-penetrate into the seabed due to pile weight, and are then embedded by evacuating seawater from the interior with a special remotely operated vehicle (ROV)-mounted pump skid [1,5,6].

Suction anchors are normally transported offshore on a large AHV. If the hired AHV does not have the deck space to accommodate them, the suction anchors can be brought to the installation site on a transportation barge. In the latter case, the AHV needs to come alongside the barge to pick up the suction anchors with its crane, which is typically used to lower the anchor to the sea bed, as shown Fig. 11.1. The crane stops lowering the anchor at a few meters above the seabed, as shown in Fig. 11.2. An ROV is used to monitor the anchor position and orientation to satisfy the allowable tolerance. At the correct position and orientation, the crane carefully controls the lowering action to allow a self-penetration of the suction anchor into the seabed. The anchor penetrates under its own weight to an initial depth.

**FIGURE 11.1**   Suction pile anchor is lowered into the water by the crane of the AHV. *AHV,* Anchor handling vessel.

After self-penetration is completed, an ROV pump is installed onto the anchor top. Further penetration to the final depth is accomplished by closing the evacuation valves and pumping seawater out of the anchor interior to create suction. The anchor orientation, inclination, and penetration need to be continuously monitored. Having achieved the required penetration depth, the suction pump is disconnected and the butterfly valve on the anchor top is shut. The installation of the suction anchor is complete.

Driven pile anchors are installed in a similar manner as suction pile anchors. They are lowered to the sea floor by a crane. The pile penetrates to an initial depth under its own weight. Penetration to final depth is

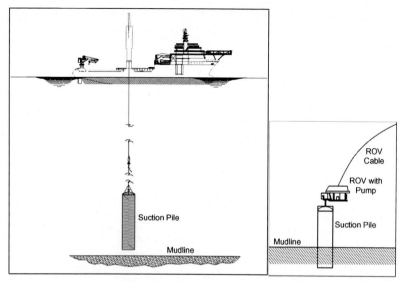

**FIGURE 11.2** (Left) Suction pile anchor is lowered to a certain height above seabed ready for self penetration. (Right) ROV pumps seawater out of the anchor interior to create suction for further penetration. *ROV*, Remotely operated vehicle.

accomplished by the use of a pile hammer mounted on the pile top. Note that the operating depth of hydraulic hammers is normally limited to about 5000 ft of water depth [6]. Alternatively, the pile can be drilled and grouted in place, or the pile can be dropped from a calculated height above the sea floor using gravity to reach the design penetration depth.

## 11.2.2    Phase II—prelay of mooring lines on seabed

Before starting on prelaying a line, an ROV survey is performed to search for obstructions that could interfere with the work along planned mooring line prelay routes and the anchor locations. The sea floor survey ensures that a prelay corridor for each mooring line is free of obstructions.

If the mooring lines have polyester segments, the polyester ropes are spooled from its storage reel to one of the winch drums on the AHV. Once the polyester ropes are ready on the winch drum, the bottom chain segment can be overboarded and lowered. At the lower end of the bottom chain is a subsea connector, as shown in Figs. 11.3 and Fig. 11.4. Mooring components are connected on the deck of the AHV (as shown in Fig. 11.5) and deployed one by one according to the procedure. Polyester rope will follow the bottom chain, and gets overboarded and deployed. Having paid out one polyester rope segment, the lowering stops and the next rope segment is connected via a connector (e.g., H-link). When the male subsea connector eventually gets lowered to the top of the suction anchor, an ROV is used to connect the

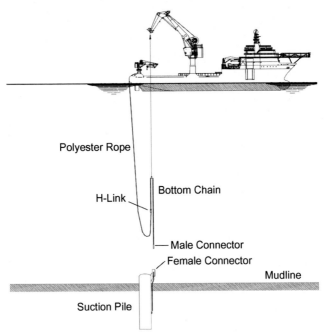

**FIGURE 11.3**   Lowering the bottom chain for connection to forerunner chain on the anchor pile.

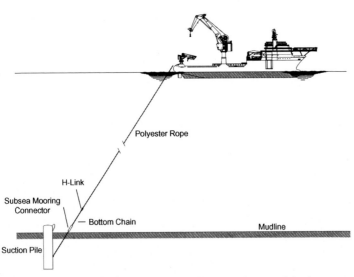

**FIGURE 11.4**   AHV is ready to prelay mooring line once bottom chain is connected to anchor's forerunner chain. *AHV*, Anchor handling vessel.

**FIGURE 11.5**    Connecting polyester rope with an H-link to bottom chain.

male to the female connectors. The female connector is temporarily seated on the top of the suction anchor and sealed with a cap that is removed by an ROV, prior to stabbing by the male subsea connector. Refer to Section 9.6 for more details about connectors.

Once the bottom chain is connected to the forerunner chain as shown in Fig. 11.4, the mooring line can be slowly paid out and laid on the seabed while the AHV moves along the prelay route. The procedure is repeated for the rest of the mooring legs. The prelay operation is complete, and the mooring lines are wet-parked on the seabed waiting for the hook-up operation in the next phase.

Sheathed wire ropes can be wet-parked as long as they are carefully laid to avoid bending and compression. They are typically prelaid as straight segments. However, note that polyester ropes can be wet-parked only if they have a qualified design that utilizes layers of cloth filters inside the rope jacket to resist soil ingression. In 2009 the US regulatory agency, Mineral Management Service, issued a Notice to Leaseholders, which allowed not only MODUs but also permanent polyester systems to be prelaid on the seabed provided that certain criteria are met [7]. The ability to prelay polyester ropes on the seabed offers advantages to conventional polyester mooring installation methods that involve suspending the polyester from either a surface or submerged buoy. Surface buoys tend to break away because the shackle and padeye at their bottom can suffer from wear and fatigue in waves. Submerged buoys make the installation procedure more complex. Prelaying polyester ropes on the seabed eliminates these and many other problems [4] and is becoming the preferred method in the industry.

### 11.2.3 Phase III—hook-up of mooring lines to floating production unit

The prelaid mooring lines may be sitting on the seabed for a period of time, say from a couple of months to 1 year, until the hull is constructed and finally towed to the site. Getting ready for the hook-up, the FPU hull is kept in position by a few (e.g., three or more) towing tugs. Two tugs connect to one side of the FPU, and one or two tugs to the other side. Once the floating hull is towed to the site and the weather condition is within the allowable limit, the hook-up procedures of the mooring lines can commence.

First, the prelaid mooring line is picked up from seabed by the AHV, as shown in Fig. 11.6. Then, platform chain is connected to it on the deck of the AHV. The AHV approaches the FPU with its stern toward the FPU.

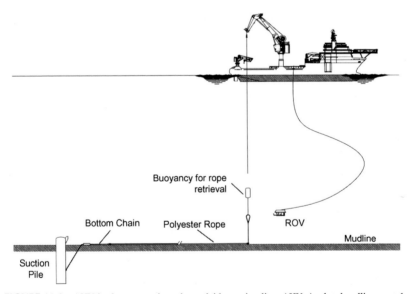

**FIGURE 11.6** AHV is about to retrieve the prelaid mooring line. *AHV*, Anchor handling vessel.

Work (pennant) wires are used as an aid to hand over the installation chain (temporary work chain) from the FPU to the AHV. The pennant wire is transferred to AHV by FPU's crane. With that, the AHV can pull the installation chain onto its deck until it can be secured in the shark jaw, as shown in Fig. 11.7. While the FPU pays out the installation chain, the AHV moves slightly away from the FPU, pulling the end of platform chain on deck and secures it by the other shark jaw. Platform chain and installation chain are both on the back deck, as shown in Fig. 11.8. The two chains can now be connected by a special connecting link, i.e., LLLC link. Note that an LLLC link is designed to pass through fairleads and chain jacks like a common link.

**FIGURE 11.7**  Chain gets pulled through the towing pins (left) on the deck and secured by the shark jaws (middle).

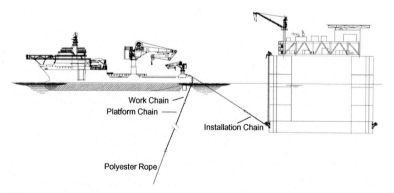

**FIGURE 11.8**  Platform and installation chains are brought to the AHV's deck for connection. *AHV*, Anchor handling vessel.

As the FPU pulls in the installation and platform chains using its winching equipment (e.g., chain jack), the AHV pays out a work wire to lower and release the mooring line (as shown in Fig. 11.9). The AHV is released from the installed line by ROV cutting the sacrificial wire sling. Final pull-in and tensioning will be completed by the FPU's chain jacks (as shown in Fig. 11.10). The same procedure described above is repeated for the other three corners of the semisubmersible. Once a specified number of mooring lines, such as four lines, are hooked up, the FPU reaches a condition called "storm safe." The partially installed mooring system has gained a limited capability of station-keeping to resist a storm of a certain level. The procedure is then repeated for the rest of the mooring legs, and the main procedure of the hook-up is complete.

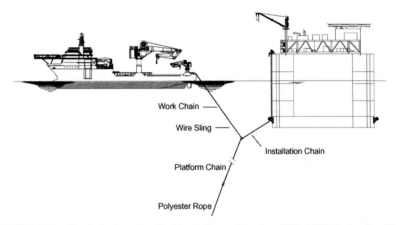

**FIGURE 11.9** The hooked-up mooring line is lowered by the AHV. *AHV,* Anchor handling vessel.

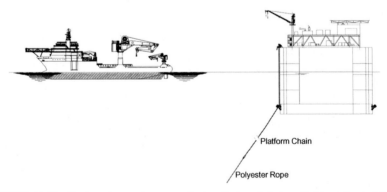

**FIGURE 11.10** Hook-up is complete, once the installation chain is pulled in and stored in a chain locker.

If the mooring line has any polyester rope segment, an extra amount of tension is intentionally applied to remove the "construction stretch" in the polyester ropes [4]. Right after the hook-up, polyester ropes are typically tensioned to about 40% of the rope minimum breaking load (MBL) for 2 hours to remove the construction stretch. Refer to Section 9.4.3 for more details on polyester stretch. Because the chain jacks are normally not designed to have the extra tensioning capacity, a cross-tensioning technique can be used to stretch the polyester ropes. The technique utilizes two chain jacks at one corner to pull in order to tension up one mooring line at the opposite (cross) corner of the FPU. This two-to-one crossing tensioning technique allows a polyester rope to be pulled at a high tension, such as ~40% of its MBL. Once the construction stretches are removed according to the predefined procedure, the tensions are lowered to the desired pretension.

Following the completion of the mooring hook-up, a visual postinstallation survey of the mooring system is performed by an ROV. The survey documents the as-laid configuration of the mooring lines, notes any twist in lines, and looks for any damage introduced during installation. It summarizes the pretensions, line angles, and positions of the installed mooring components. The video footage of the ROV survey is recorded and archived. These data will serve as the baseline for the integrity management of the mooring system throughout the field life. At this point in time, the mooring installation is considered fully complete, and the installation of risers, umbilicals, and other equipment may commence.

## 11.3    Deployment and retrieval of temporary mooring

Installation procedures for temporary moorings are different from those permanent moorings. For temporary moorings that do not use a preset mooring (see Section 11.3.2), equipment such as anchors and wire ropes are carried by the subject vessel which could be a MODU, floater, construction/work barge, or tender assisted drilling. An AHV is used to deploy the mooring equipment. It needs to be understood that the subject vessel will move, after a few weeks or months, to another site whenever the intended operation is complete. The deployed mooring legs have to be retrieved and brought back to the subject vessel with help from an AHV. Therefore the installation of a temporary mooring system is often phrased as "deployment and retrieval" rather than installation.

For temporary moorings, a site-specific geotechnical survey is normally not conducted. Best available existing geotechnical data from the surrounding areas or nearby fields are used to assess the suitability and fitness of the rig anchor. If detailed site-specific soil data are available for a mooring location, this information should be used. However, such information may not be available for some exploration wells that could be in a frontier (new) area. In that case, geotechnical and geophysical surveys can be essential. An ROV survey in mooring line corridors to check for geohazards and constraints is normally conducted before the mooring deployment.

### 11.3.1 Rig mooring system for mobile offshore drilling unit

MODUs are generally moored with 8 to 12 anchors. Mooring lines are laid in a spread pattern. The deployment is conducted by AHVs that have large engine power to handle rig chain, wire, and anchors. Fig. 11.11 illustrates a MODU that is assisted by two AHVs to deploy its eight anchors. The newer generation of AHVs has bollard pulls in excess of 300 tons or higher. The AHV needs to be properly selected to ensure that it has the capability to handle the rig chain and anchors.

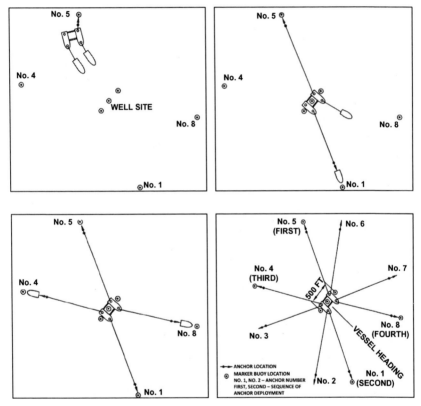

**FIGURE 11.11** A MODU is assisted by two AHVs to deploy its rig anchors. *AHV*, Anchor handling vessel; *MODU*, mobile offshore drilling unit.

The typical method for deployment and retrieval of rig anchors on MODUs is to use a "chaser" from the AHV. A chaser is a ring-shaped or hook-shaped tool that is used to chase (slide) along the mooring line toward the anchor and back again to a rig or handling vessel. Its function is to grab and move an anchor during a deployment or retrieval operation. Besides using a chaser, anchors may be handled using a pendant and buoy. An anchor pendant is a wire that is attached to the crown of an anchor enabling the anchor to be pulled out of the seabed.

To deploy a rig mooring system, an AHV unracks (removes) a rig anchor from the MODU's bolster, as shown in Fig. 11.12. The AHV then runs the anchor line out the full distance to the anchor location with the anchor on the deck or on the roller. The AHV increases power until anchor line tension rises on the MODU winch tension meter. The anchor is overboarded and lowered over the stern roller. Note that the anchor needs to stay correctly oriented in the chaser at all times. The AHV lays the anchor on the seabed. The MODU pulls in (heave in) the rig wire rope to drag and set the anchor, as shown in Fig. 11.13. The embedment of rig anchors is obtained by dragging. Unlike pile anchors, a load test is required for drag anchors to assure that the holding capacity is achieved to the desired level. The AHV can retrieve the chaser and return to the MODU to deploy the next anchor and mooring line.

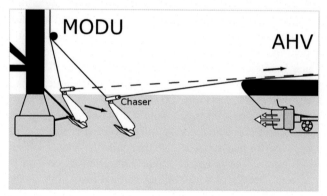

**FIGURE 11.12**   Rig anchor is unracked from MODU's bolster (rack) for deployment by AHV. *AHV*, Anchor handling vessel; *MODU*, mobile offshore drilling unit. *Courtesy Vryhof.*

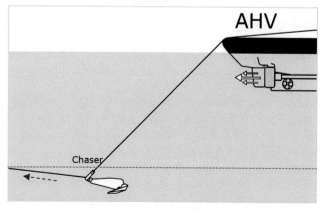

**FIGURE 11.13**   Rig anchor is getting embedded in seabed while handled by AHV. *AHV*, Anchor handling vessel. *Courtesy Vryhof.*

## 11.3.2 Preset mooring system for mobile offshore drilling unit

Preset moorings have been commonly used with permanent mooring applications for FPUs. They can also be applied to temporary moorings. A preset mooring is a system where the majority of off-vessel components are installed prior to arrival of a MODU on location. The components in a preset system may include wire ropes, chain, polyester ropes, and underwater buoys. The preset components are typically buoyed off, and sequentially hooked up to the MODU rig using an AHV. The existing chain or wire onboard the MODU (i.e., rig chain and rig wire) are simply connected to the preset moorings.

With preset moorings, MODUs of older generations can considerably extend water depth limits [8], which is their main advantage. Polyester rope, due to its light weight, can substantially help such an extension to deeper water. Generally, only one AHV is required for the preset mooring installation work. The main requirement for the preset AHV would be the provision of a large main winch and considerable storage capacity for storing wire and chain.

Another advantage of using a preset mooring is that the MODU can come to the site and hook up quickly, so that the drilling operation can be started sooner. In other words, a preset mooring can increase the drilling uptime, which is often a priority. With a preset mooring, the disconnect and reconnect operations are simple, less likely to have complications, and have less potential for weather downtime [8].

The use of taut leg moorings can benefit a preset system by reducing the amount and scope of components. A taut leg mooring typically gives a better station-keeping performance, as it can maintain the vessel offset within a smaller watch circle. The issue for a taut leg system is how to provide an anchoring system capable of withstanding vertical loads. Pile anchors, either driven or suction, have the capability to resist vertical loads, but the cost of installation and the inability to reuse them makes piles unattractive for temporary moorings. Driven piles also have a water depth limit of about 5000 ft, which is constrained by the current underwater hammer capability [8]. Preset moorings with pile anchors can also solve the problem of a congested field where anchor locations and drag distances are limited by subsea equipment such as pipelines [1,9]. Vertically loaded anchors (VLAs), which are a special type of drag embedment anchor, are another solution and a good alternative to piles for taut leg moorings.

Because of the demand for deepwater-capable MODUs, drilling operators may consider the preset mooring option. For a short-term drilling contract of one or two wells, it would be difficult to justify the initial capital cost of the preset system. For a contract of 1 year or longer, the overall economics may become competitive compared to a conventionally moored MODU of a newer generation or a dynamic positioning MODU. In summary, preset moorings can be used for both old and new MODUs for deepwater drilling and exploration. Polyester rope and VLAs make preset moorings practical and viable.

## 11.4    Installation vessel

Mooring installations are normally performed by AHVs, but they can also be done by other types of offshore vessels such as construction barges. However, the latter can be significantly more expensive to lease than the former.

### 11.4.1    Anchor handling vessel

An AHV is an offshore supply vessel specially designed to provide anchor handling services and to tow offshore platforms, barges, and production vessels. AHV is also known as anchor handling tug (AHT). As the main type of offshore supply vessels, AHVs have been used mainly for offshore drilling and production activities. They serve multiple purposes including the following:

- handling anchors and mooring lines for drilling rigs or production units;
- towing of floating structures in open waters with subsequent positioning on site;
- deploying subsea equipment; and
- providing supply services.

AHVs can also be used as standby rescue vessels for oilfields in production. They are often used in general supply service for all kinds of platforms, transporting both wet and dry cargo in addition to deck cargo. While they may be referred to as tugs, they are not designed to push other vessels alongside like a harbor tug boat.

Large AHVs can have a bollard pull over 300 metric tons. The required bollard pull of an AHV has a major influence on the vessel design, since it defines the power need, the propeller size, hull shape, and depth aft to give the necessary propeller immersion [10]. Their hull beam and shape are designed such that they provide good stability, particularly when heavy moorings and anchors are suspended from the stern. Anchor handling requires high power, winch capacity, deck space, storage lockers for rig chains and auxiliary handling equipment. A stern roller is used to ease the passage of wires and anchors over the stern of the vessel during deployment of the anchor. Fig. 11.14 shows the general layout and major equipment of a large AHV.

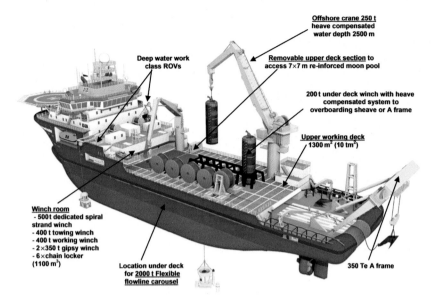

**FIGURE 11.14** AHV with a large crane, an A-frame, and two ROVs. *AHV*, Anchor handling vessel; *ROV*, remotely operated vehicle. *Courtesy SBM Offshore.*

The demand for AHVs has been up and down in the oilfield market. The years from 2006 to 2008 witnessed the booming period for the offshore support vessel industry. In 2007 new orders for AHVs hit a record high of 362 vessels, while the figure dropped to 201 units in 2008 due to oversupply. New AHV orders saw a sharp decline to roughly 56 units in 2011. As the oilfield market moves with the oil price, the demand for AHVs fluctuates as well. As the oil industry migrates into deeper water, those AHVs with a large main winch, large storage capacity (e.g., in chain lockers), large deck space (to stow anchors as shown in Fig. 11.15), and high bollard pull may stay in demand.

### 11.4.2 Anchor handling vessel incident—capsizing of Bourbon Dolphin

On April 12, 2007 the AHV "Bourbon Dolphin" was engaged in anchor handling operations for a semisubmersible drilling rig west of the Shetland Islands. An adverse weather condition and the weight of the mooring chain hanging from her stern put Bourbon Dolphin in a severe list to the port side. The vessel rolled to a large angle and capsized. Seven of the 14 crew members were lost [11].

The investigation report [11] identified 10 factors that contributed to the root cause of the incident, and then summarized: *It is clear to the Commission that it was particularly the changed angle of attack from*

**FIGURE 11.15**   Large deck space on an AHV allows more mooring components to be carried. *AHV*, Anchor handling vessel.

*the chain, together with the vessel's heading and reduced maneuverability, plus influence from external forces, that, together with the vessel's stability characteristics and current load condition, made the accident possible.*

In short, the problem in the incident was that the vessel stability was detrimentally compromised by several factors. The heavy rig chain hanging from the port-side (left-side) towing pin is one of the main factors that forced the AHV to list and capsize. The incident clearly shows that mooring installations can be a dangerous job. Vessel stability of any floating systems is a critical discipline that should be well understood and managed by responsible parties. Careful planning, preparation, and personnel training are crucial to ensure a safe operation in the offshore environment.

## 11.5   Questions

1. What is the purpose of a geophysical survey? And, what is the purpose of a geotechnical survey?
2. What can be the three phases of a mooring installation for a FPU?
3. Polyester fibers are very susceptible to abrasion or cuts. Can polyester mooring ropes be parked on seabed for a period of time before the hookup to a floating structure? How is the concern addressed?

**4.** Give at least one reason why a drilling semisubmersible can benefit from a preset mooring system.
**5.** List at least two factors that contributed to the capsizing of the AHV, Bourbon Dolphin.

## References

[1] API RP-2SK, Design and Analysis of Stationkeeping Systems for Floating Structures, third ed., 2005.

[2] Y. Bai, Q. Bai, Subsea Engineering Handbook, Gulf Professional Publishing, Elsevier, 2010. ISBN: 978-1-85617-689-7.

[3] ABS, Guidance Notes on Design and Installation of Drag Anchors and Plate Anchors, American Bureau of Shipping, 2017, updated 2018.

[4] T. Veselis, R. Frazer, M. Huntley, Mooring design and installation considerations for the mirage and telemark field development, in: OTC 21018, OTC Conference, May 2010.

[5] DNV-RP-E303, Geotechnical design and installation of suction anchors in clay, in: Recommended Practice, DNV-RP-E303, 2005.

[6] I.J. Witchers, Guide to Single Point Moorings, July 2013.

[7] MMS, NTL No. 2009-G3, Synthetic Mooring Systems, January 2009.

[8] P. Dove, T. Fulton, Pre-Set Moorings Provide Less Costly Alternative to DP in Ultra-Deepwater, Offshore Magazine, 1997.

[9] K. Ma, R. Garrity, K. Longridge, H. Shu, A. Yao, T. Kwan, Improving reliability of MODU mooring systems through better design standards and practices, in: OTC 27697, OTC Conference, May 2017, 2017.

[10] Wartsila, Wärtsilä Encyclopedia of Marine Technology. <www.wartsila.com>, 2018.

[11] NOU Official Norwegian Reports, The Loss of the Bourbon Dolphin on 12 April 2007, Government Administration Services Information Management, Oslo, 2008, ISSN 0333-2306, ISBN 978-82-583-0965-6.

# Chapter 12

# Inspection and monitoring

## Chapter Outline

12.1 **Inspection** **233**
    12.1.1 Regulatory requirements 234
12.2 **Inspection schedule** **234**
    12.2.1 As-built survey for permanent mooring 235
    12.2.2 Periodic surveys for permanent mooring 235
    12.2.3 Periodic surveys for Mobile Offshore Drilling Unit mooring 236
12.3 **Inspection methods** **236**
    12.3.1 Difference between Mobile Offshore Drilling Unit and permanent moorings 236
    12.3.2 General visual inspection 237
    12.3.3 Close-up visual inspection 239
    12.3.4 Nondestructive examination techniques 240
    12.3.5 Advanced three-dimensional imaging 240
12.4 **Inspection of mooring components** **241**
    12.4.1 Inspection of chain 241

12.4.2 Inspection of wire rope 243
12.4.3 Inspection of fiber rope 244
12.4.4 Inspection of connecter and anchor 245
12.5 **Monitoring** **245**
    12.5.1 Regulatory requirements 246
    12.5.2 What and how to monitor 246
12.6 **Monitoring methods** **247**
    12.6.1 Method 1—monitoring visually 247
    12.6.2 Method 2—monitoring tension 248
    12.6.3 Method 3—monitoring vessel position 248
12.7 **Monitoring devices** **249**
    12.7.1 Load cell 250
    12.7.2 Inclinometer 250
    12.7.3 Global Positioning System—based system 251
12.8 **Questions** **252**
**References** **252**

Inspection plays a critical role in the asset management for owners and operators, as good inspection practices prevent mooring incidents caused by poor condition of mooring components. This chapter summarizes current inspection practices and methods. It also provides guidance on mooring system monitoring.

## 12.1 Inspection

Mooring components are designed to allow for limited degradation, such as wear and corrosion, so inspection is necessary to confirm that the

Mooring System Engineering for Offshore Structures. DOI: https://doi.org/10.1016/B978-0-12-818551-3.00012-0
**233**

degradation is within the design limits. In addition, inspection is performed to monitor the integrity by discovering anomalies in individual components.

### 12.1.1   Regulatory requirements

Most floating structures follow the framework of in-service inspections defined in Class Rules. The framework is based on long-term practices established by the shipping industry. Periodic surveys are performed on the hull and mooring systems, which can be classified as Annual, Intermediate, and Special Surveys. Note however that the owners can choose not to class the floating structures and can define their own framework of in-service inspections.

While Class Rules define the inspection schedule and scope, they provide limited guidance on potential damage modes for each type of mooring component. American Petroleum Institute Recommended Practice (API RP) 2I is one of the standards that gives comprehensive guidance and explicit discard criteria specifically for mooring components. The first edition of RP 2I [1] was originally written for the inspection of Mobile Offshore Drilling Unit (MODU) moorings, and largely relied on "in-air" inspections. The third edition published in 2008 [2] incorporated recommended practices for the inspections of permanent mooring systems and provided guidance on underwater inspection techniques. RP 2I is prescriptive and gives discard criteria for certain types of defects, e.g., cracks on chain and broken wires on wire rope. It also specifies what inspection intervals should be used and the extent of the inspection for some types of mooring components.

The philosophy of API RP 2I is to prevent excessive deterioration of the mooring components from the original condition. Based on this philosophy, a criterion of allowing a strength reduction of up to 10% Minimum Breaking Strength was established in the first edition, which is primarily specified for MODU mooring components. While this 10% criterion has been widely used for more than 20 years, it is considered controversial by some mooring engineers due to its lack of engineering basis. They argued that a check against the design safety factors should be the criterion. In other words, the mooring component is allowed to stay in service, only if a design check proves that the mooring component has a sufficient amount of remaining capacity to meet the safety factor. The controversy is yet to be resolved.

## 12.2   Inspection schedule

Inspection of a permanent mooring system has two stages, i.e., as-built and in-service. The first occurs after mooring hook-up when an as-built survey is conducted. After that, mooring systems are inspected periodically at various levels of detail based on Class or owner's requirements. In addition, some operators may inspect the mooring system after severe storms or other events that warrant inspection. MODU moorings are quite different from permanent moorings, as the former can be inspected whenever the components are retrieved. The good accessibility allows MODU mooring components to be closely examined in a dry condition.

## 12.2.1   As-built survey for permanent mooring

An as-built survey should be performed once the mooring system is hooked-up to the floater and tensioned to the design values. The survey is primarily conducted to confirm that the anchor legs are connected as designed, to check for damages that occurred during installation, and to ensure that the twist in the anchor legs is within the design margins. Any discrepancy in tension or angle from design values should be addressed. The as-built survey also serves as the baseline for comparison with all subsequent inspections over the service life. The inspection should be documented accurately with sufficient detail.

Most as-built surveys are conducted from the anchor to the fairlead or as close to the water surface as practical, and are primarily visual inspections performed with a remotely operated vehicle (ROV). The visual inspection is usually video recorded along with comments made by the inspector. In many cases the ROV position and depth can be recorded. The as-built details should be documented in order to facilitate future inspections. The as-built report should include a detailed listing of all components in each mooring leg such as manufacturer, serial number or other identification.

## 12.2.2   Periodic surveys for permanent mooring

To ensure that a mooring system behaves as designed, it is necessary to perform inspections on a regular basis to monitor and remedy the condition of the various components. Visual inspections tend to be performed on a yearly basis for components that are easy to access, and once every 5 years, as a minimum, for other components. Inspection opportunities are limited, particularly in the touch down zone and at underwater vessel connections. Furthermore, it is generally impossible to see the buried chain between the dip point and the anchor.

Some of the owners' inspection plans follow API RP 2I [2]. Many adopt the framework defined in Class Rules [3–5], which can be classified as follows:

- *Annual survey*—Mooring components above the waterline are inspected on an annual basis. Attention should be paid to chain in contact with any winches, chain stoppers/fairleads, and in way of the splash zone.

- *Intermediate survey*—Intermediate survey may or may not occur every 2.5 years, depending on which Class Society is used. If occurring, it is typically conducted at the 2nd or 3rd annual survey. It may take the form of an "in-water" survey. The premise of an in-water survey is to provide the same information normally obtained from a docking survey as far as can be practically achieved.
- *Special survey*—Special survey occurs every 5 years. Where possible, the mooring system equipment should be raised to the surface for detailed inspection. Alternatively, in situ (in-water survey) inspection can be adopted. The scope of this inspection should include the annual inspection requirements as well as mooring components at or near the touch down point, any prior-noted damage to the mooring system, determination of the extent of marine growth, condition and performance of any corrosion protection system fitted.

Some owners' inspection plans might specify an additional ROV survey in addition to the scopes defined by Class. This occurs especially when the mooring design incorporates certain technology, such as polyester ropes. A nonperiodic inspection may be conducted by the owner, if an event warrants an inspection, for example, a passing vessel is known to drag a work wire across the mooring footprint. Unplanned inspections can also occur after passing of a storm that exceeds design conditions.

### 12.2.3 Periodic surveys for Mobile Offshore Drilling Unit mooring

Inspection schedule for MODU chain or wire rope follows roughly the framework defined in Class Rules, but is also based on the age, condition, and operational history of the mooring component and type of operation. Some recommended inspection intervals for chain and wire ropes can be found in API RP 2I. The inspection intervals may be modified based on the actual condition and previous inspection record of the mooring component.

## 12.3 Inspection methods

### 12.3.1 Difference between Mobile Offshore Drilling Unit and permanent moorings

One of the approaches for MODU mooring inspection is shown in Fig. 12.1. The drilling vessel is taken into a dock, and the chain is laid out on a dry surface for inspection. Normally such chain inspection is carried out in conjunction with other work such as major structural repair or special survey. In this manner, the entire chain can be thoroughly cleaned and carefully inspected, and the connecting links and anchor shackles can be examined by magnetic particle inspection (MPI).

**FIGURE 12.1** Dockside inspection method for MODU chain. *MODU*, Mobile Offshore Drilling Unit. *Courtesy: Vryhof.*

In another approach, the drilling vessel stays offshore, and the chain is inspected with the assistance of a workboat. The chain in the chain locker is paid out fully and then examined by an inspector standing close to the windlass while the chain is slowly taken back into the chain locker. At the same time, the workboat picks up the anchor and moves slowly toward the vessel. The advantage of this method is that it requires no dock facilities. The inspection can be performed whenever a workboat is available or in conjunction with anchor retrieval. Similar arrangements can be used to inspect mooring wire ropes.

Unlike the MODU moorings which can be inspected in-air during anchor deployment or retrieval, permanent moorings are inspected in-place. The above water components can be inspected visually or by nondestructive techniques such as MPI. The underwater components can be inspected by divers in shallow water or ROVs in deep water. The inspection is mainly visual with photographs, video, and inspectors' comments taken. Direct measurement of component diameter or size, if possible, provides quantitative data for component assessment. Depth measurement of connectors can provide useful feedback on overall mooring system performance. If the anchor leg system is provided with a load monitoring system or fairlead angle indicator (inclinometer), data can be recorded along with floater position to enable accurate assessment of the mooring line tensions.

## 12.3.2    General visual inspection

General visual inspection is the most common method that is carried out by a continuous slow ROV flight or diver swim past the items being inspected. It is used to assess the overall condition of the mooring legs and to determine if any further inspection is required. It relies on having sufficient water clarity, adequate camera quality, and the knowledge of the inspector. Although video recording is normally made for later review, it is more effective for an inspector to observe in real time and instruct the repositioning of the ROV to get views from different angles. Some inspectors have considerable experience with mooring equipment inspections. However, many of them cover a variety of inspection scopes, including hulls, risers, and subsea equipment, and may not be specialized in moorings. The knowledge level of an inspector on moorings can be a major factor in being able to identify anomalies.

The most common tool for inspecting mooring lines is the *ROV*. The inspections are performed from a dynamic positioning (DP) vessel that is suitably equipped for the operation. The vessel will maneuver around the mooring leg to allow the ROV to fly down one side of the leg and then fly back up the other side, so a complete visual observation and record of the inspection is completed. The inspections allow external damage to be found and could indicate some forms of internal damage. Many ROVs are large work-class ROVs operating from either the floating system or more likely from a dedicated vessel. Videotaping capability is a requirement that will allow a complete real-time record of the inspection process. Fig. 12.2 shows video images of a mooring wire rope taken by an ROV during an inspection.

**FIGURE 12.2** Underwater inspection of a wire rope by an ROV [6]. *ROV*, Remotely operated vehicle.

In shallow waters, *diving* has a successful history, but with considerable safety risks to personnel. Divers, whether near surface or using saturation techniques, require all the correct precautionary equipment, and typically this would come with a dedicated dive vessel; hence any marine risks need to be considered from multiple vessels in the field working within a vessel swing circle. Nonetheless, diver inspections are in general not a favored option. Mooring lines are highly dynamic, and therefore are potentially dangerous when divers are in close proximity. Also, diver inspection has inherent depth limitations.

### 12.3.3   Close-up visual inspection

Close-up visual inspection is directed at a particular mooring component or item. Its purpose is to assess the condition of the subject component and to measure any anomaly. It requires access such that the item under scrutiny is within arm's reach of the inspector or can be closely viewed by the ROV. Cleaning of the area will normally be required. Fig. 12.3 shows a mooring chain covered by heavy marine growth and the same chain after cleaning by a diver. Figs. 12.4 and 12.5 show rope-access specialists taking diameter measurements using a caliper on mooring chain under an external turret.

**FIGURE 12.3**   Marine-fouled chains before and after cleaning by a diver [7,8].

**FIGURE 12.4**   Rope-access specialists climbing down to inspect chain in splash zone.

**FIGURE 12.5** A rope-access specialist measures chain diameter using a caliper [9].

Visual inspection at a close distance can be risky. Not all ROVs are capable of flying sufficiently close to a mooring component. For a diver, there may be too great a risk of unexpected motions to approach sufficiently close. For example, when inspecting a chain inside a hawse pipe (trumpet) at the bottom of a turret, large work-class ROVs are unable to get in close enough, nor able to bring appropriate lights into the trumpet mouth to examine suspected wear.

### 12.3.4 Nondestructive examination techniques

Nondestructive examination (NDE) techniques refer to those for identifying surface crack. They are performed on suspect components that are identified by visual inspection. They are also used on critical locations identified by analysis or in-service performance experience. The NDE techniques most commonly used are (1) MPI (magnetic particle inspection) or DP (dye penetrant) where no coating is present; and (2) eddy current inspection for the detection of surface breaking indications through paint coatings. In the case of fatigue crack detection, MPI and DP are considered the most appropriate methods when the components are dry in air.

### 12.3.5 Advanced three-dimensional imaging

A three-dimensional imaging system is a high-tech inspection tool that has become mature in recent years. It can be very useful for examining problem areas, particularly for identifying sizes of abrasion areas or corrosion pits in chain links. A number of companies offer this special capability, which can be added to most ROVs. However, its full potential is only possible after thorough cleaning of the mooring components, and then accurate geometries can be created for postprocessing. One particular system has shown successful application. It is comprised of multiple high-resolution video cameras and lights on a deployment frame. The system measures the chain parameters by calibrating with a tool and resolving dimensions with offline image-analysis software. This optical chain measurement technology has been used occasionally by some operators to assess mooring components in critical (marginal) conditions.

## 12.4    Inspection of mooring components

Along the length of a mooring line, there are a few areas that are more prone to integrity issues [10,11]. These areas deserve special attention during an inspection. The top end at vessel interface and the touch down area at the seabed may be the two most problematic areas. They are subject to high degradation and should be most closely inspected. The splash zone is another inspection focus area due to potentially severe chain corrosion. All connectors and wire rope terminations are also critical components, because the discontinuity in weight per length can cause increased relative bending and wear. In this section, a brief guide is provided for each component type and what to look for during an inspection.

### 12.4.1    Inspection of chain

The rough environment to which mooring chain is exposed can lead to various chain problems, as discussed below.

*Corrosion*—General corrosion is commonly seen on mooring chain in the splash zone, as shown in the top two photos of Fig. 12.6. It can be very aggressive with a corrosion rate higher than 1.0 mm/year depending on the quality and temperature of the seawater. Also, large pits can develop in submerged chain mostly in the upper water column, as shown in the bottom two photos in Fig. 12.6. These pits are presumably caused by sulfate reducing bacteria and this is a key contributor to MIC (microbiologically influenced corrosion) [13]. Excessive corrosion increases the possibility of chain failure from fatigue or overloading due to the surface feature and the reduced cross-sectional area.

*Wear or abrasion*—Wear between links and the wildcat of fairlead, or between two adjacent links reduces the chain diameter. The diameter reduction decreases the load-carrying capacity of the chain and may invite failure. Chain wear can also happen in the touch down area. Fig. 12.7 shows the material loss on one side of touch down/ground chain with a noticeable flattening. It is suspected that these are cause by either MIC or seabed abrasion.

**FIGURE 12.6**    Chain with heavy general corrosion (top two); chain with large corrosion pits (bottom two) [7,8,12,13].

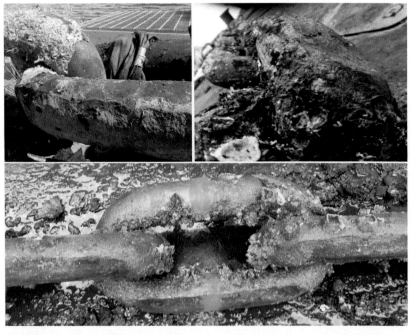

**FIGURE 12.7**    Ground chain and touch down chain with material loss on one side [9,12].

*Cracks*—Surface cracks, flash-weld cracks, and stud-weld cracks may propagate under cyclic loading, resulting in premature chain failure. Note, however, these surface cracks are nearly impossible to be detected by the naked eye, even after the marine growth is thoroughly removed.

*Loose or missing studs*—A stud chain link without a stud may result in higher bending stresses and lower fatigue life for the chain. A loose stud in a stud link caused by abusive handling or by excessive corrosion between the link and the stud allows excessive stretching of chain, causing higher bending stresses in the chain.

*Gouges*—Physical damage to the chain surfaces such as cuts, pits, and gouges raises stress and may promote fatigue failure.

*Elongation*—Excessive permanent elongation may cause a MODU chain to function improperly in the wildcat, resulting in bending and wear of the links. Wear in the grip area of the chain and working loads in excess of the original proof load will result in a permanent elongation of the chain.

### 12.4.2    Inspection of wire rope

Mooring wire ropes can experience various types of damage as discussed below.

*Broken wires*—Broken wires at the termination (Fig. 12.8), even if few in number, indicate high stresses at the termination and may be caused by incorrect fitting of the termination, fatigue, overloading, or mishandling during deployment or retrieval. If broken wires are closely grouped in a single strand or adjacent strands, there may have been local damage at this point. When wire breakage of this type begins, it will usually only get worse. Such concentrated wire breakage will upset the balance of loads carried by the strands.

**FIGURE 12.8**   Broken wires at wire rope termination [14].

*Corrosion*—Corrosion in the marine atmosphere not only decreases the breaking strength by reducing the metallic area of the rope, but also accelerates fatigue by causing an irregular surface that will invite stress-cracking.

Severe corrosion may reduce a rope's elasticity. Corrosion of the outer wires is a common problem and may be detected visually. Internal corrosion is more difficult to detect than the external corrosion that frequently accompanies it.

*Loss of lubrication (blocking compound)*—Proper and thorough lubrication (blocking compound) is important to permit the wires and strands to work without excessive internal wear and to inhibit corrosion. Operating a wire rope in frequent bending service without lubrication will reduce its life to only a fraction of normal life because of internal wear.

*Wear*—Wear of the crown wires of outer strands in the rope can be caused by rubbing against the fairlead sheaves or hard seafloor. In particular, external wear of mooring wire ropes can be caused by dragging the wire rope on a hard seafloor during anchor deployment or retrieval. Internal wear is caused by friction between individual strands and between wires in the rope, particularly when it is subject to bending. Internal wear is usually promoted by lack of lubrication. Wear reduces the strength of wire ropes by reducing the cross-sectional area of the steel.

*Kink or deformation*—Distortion of the rope from its normal construction is termed deformation and may result in an uneven stress distribution in the rope. Kinking, bending, scrubbing, crushing, and flattening are common wire rope deformations. A kink is a deformation in the rope created by a loop that has been tightened without allowing for rotation about its axis. Unbalance of rope construction due to kinking will make a certain area of the rope disproportionately susceptible to excessive wear. Bends are angular deformations of the rope caused by external influence.

*Change in rope diameter*—The rope diameter can be reduced by external wear, interwire and interstrand wear, stretching of the rope, and corrosion. Excessive reduction in diameter can substantially reduce the strength of the rope. Therefore the diameter is typically measured and recorded periodically throughout the life of the rope. The new rope diameter is measured and recorded as a reference point.

### 12.4.3   Inspection of fiber rope

Mooring fiber ropes can experience various types of damage as discussed below.

*Cut or abrasion*—Fiber rope damage is often caused by contact of the fiber rope with sharp edges during rope deployment or retrieval. For an installed fiber mooring line, damage can also be caused by a falling object or contact with a work wire rope used for other installation activities. Damage can also be caused by external abrasion. Ropes that have worked against fixed objects or have been dragged on a hard seafloor may be subjected to damage. Damage to the splice or the jacket may occur during installation when the splice comes in contact with installation equipment. Since the jacket is not a load-carrying element in a fiber rope, minor damage to the jacket is acceptable.

*Soil ingress or marine growth*—Ingress of soil particles may occur when the rope comes in contact with the seafloor during installation, for example, a rope accidentally dropped to the seafloor. Also it may be possible for the fiber ropes of leeward mooring lines to touch the seafloor under extreme environmental conditions. To address this problem, many fiber mooring ropes are equipped with filters or soil blocking jackets that are effective in filtering soil particles. Marine growth can be harmful to fiber ropes if it penetrates through the jacket into the load-carrying fibers.

### 12.4.4 Inspection of connecter and anchor

For MODU mooring components, the inspector typically visually inspects all mooring jewelry such as anchor shackles, swivels, open links, and connecting links. In addition, certain critical areas in mooring components may be inspected by MPI. Also, the inspector visually inspects the anchors after cleaning, looking for structural cracks and noticeable deformations such as bending of the anchor shank or fluke. If a crack is suspected in an area of high stress concentration, the area is inspected by MPI.

Most permanent moorings utilize connectors such as "D" or "H" shackles, or tri-plates to connect various chain, wire rope, fiber rope, and anchor components. In addition, special subsea connectors are often used to aid in the installation of moorings and to allow rapid change out of mooring lines. Some connectors, e.g., tri-plates, can have anodes to provide corrosion protection.

One critical part in a connecting shackle is the pin with nut. It is important to ensure that the pin maintains its integrity, as there are several cases where pins have come apart due to failure of the retaining mechanism. The connectors are inspected visually for anomalies and to ensure that all retaining hardware is intact. If possible, wear measurements are taken to allow estimation of remaining strength. Corrosion can take place between the threads of the pin and the nut, so this is also inspected.

### 12.5 Monitoring

Monitoring is an important part of the asset integrity management and should be enforced as a viable means to address the condition of a mooring system, in conjunction with inspection programs. The main objective of monitoring is to continuously verify the condition or performance of the mooring system and to provide input for the assessment of mooring integrity. Operators should review how they can detect a mooring line break. It is not appropriate that a reduction in mooring capability goes undetected. Operators should specify monitoring methods or devices during the engineering phase of a project.

## 12.5.1 Regulatory requirements

Industry standards and Class Rules provide guidelines on the mooring monitoring systems, depending on the types of operation (i.e., production or drilling). As an example, API RP 2SK recommends that moored floating units should be equipped with a calibrated system for measuring mooring line tensions if the operation requires mooring line adjustment, and line tensions should be continuously displayed at each winch. For units that do not require a tension measurement device, a device for detecting mooring failure should be considered.

In general, MODUs are always equipped with line tension and vessel offset monitoring systems to meet the stringing requirements for the drilling operation. Floating production vessels are typically equipped with vessel position monitoring systems, and sometimes tension monitoring systems if the mooring lines are connected with a winching/tensioning device. For those not equipped with line tension monitoring systems, a device for detecting line failure is normally in place.

## 12.5.2 What and how to monitor

The most important parameter to observe is a failure or loss of tension in a mooring line. A failure can be detected through the loss of tension, a sudden change in line angle, a drop to the seabed, or a sudden shift in facility equilibrium position. Every floating system should assess whether it has sufficient monitoring capability to demonstrate that it is safely moored. The cost of the system is not substantial if included during construction, and monitoring can reduce the risk of pollution and production shutdown.

Monitoring can be performed through different methods, including direct measurement of *line tension*, *line angles*, or *vessel offset*. Not all methods have the same capability for line-break detection and/or tension measuring. Some of the methods require the use of batteries, which can be drained quickly depending on the amount of data collection and frequency of collection. Therefore the time interval of detection needs to be evaluated during the engineering phase based on failure consequences and the redundancy level, e.g., the number of lines in a group or the margin in factors of safety.

A *real-time* monitoring system that can provide an instantaneous indication of failure may be preferred. This is particularly the case when a mooring system has a safety factor that just barely meets the design standard and can only tolerate one broken line. One has to understand that a cascading mooring failure can happen with one break after another. The consequences of a multiline failure are potentially so high that the monitoring system must be able to detect the first break immediately to allow sufficient time to take appropriate actions. In contrast, a *nonreal-time* monitoring system may be chosen for a mooring system with a high redundancy (e.g., a larger number of lines in a group or a higher margin in safety factor) [15]. The mooring system may still be considered safely moored after the loss of one mooring leg. For such an example, a monitoring system that sends a message at longer intervals, such as a few days, may be acceptable.

Existing floating facilities that do not have a monitoring system installed may retrofit some type of monitoring device. Selection of a device among the available options needs to take into account the practicality and cost. Note that any kind of retrofit that involves an offshore installation campaign will be much more expensive compared to outfitting during the vessel construction phase.

Some monitoring methods provide a greater functionality than just detection of line breaks. Direct measurement of tension, or measurement of angle that converts to a calculated tension, may be considered an advantage in providing input data for a more in-depth analysis.

## 12.6    Monitoring methods

### 12.6.1    Method 1—monitoring visually

Visual monitoring by the crew or surveyors with (or without) a closed-circuit TV system is a common practice for an floating production storage and offloading (FPSO)/floating, storage, and offloading (FSO) with an external turret mooring system above water. The crew can take a glance at a monitor screen, as shown in Fig. 12.9, that is connected to closed-circuit cameras

**FIGURE 12.9**    Monitor screen in the control room of an FPSO showing that risers and mooring lines are intact. *FPSO*, Floating production storage and offloading.

near the bow. Alternatively, the crew can perform a daily walk-around on deck to visually confirm the existence and angle of the mooring lines. Obviously, this method is not feasible if the mooring lines are submerged underwater, such as those for internal turret moorings.

### 12.6.2 Method 2—monitoring tension

Rather than just detecting line failure, some advanced monitoring methods provide line tensions in real time. Two methods are (1) direct measurement of tension and (2) measurement of angle that leads to a calculated tension.

- *Direct tension measurement* using load cells—This is a typical setup on vessel types that use chain jacks, commonly seen on semisubmersibles and spars. Load cells are built into the chain stopper or the foundation of a winching equipment such as a chain jack sitting on deck.
- *Indirect tension measurement* using inclinometers—This is a typical arrangement for an FPSO with an internal turret. Inclinometers are fitted on the hawse pipes around the chain table at the bottom of the turret which is submerged underwater. They can also get retrofitted by clamping onto top chain. The measured angles at the top of the mooring lines can be converted to a calculated tension either from catenary calculations or look-up tables. Note that this method can introduce uncertainty since dynamic effects are not included in the calculation. Also, when the line is very slack, small changes in tension will introduce significant angle variations; however, while the line gets tauter, the angle changes become less significant. Still, inclinometers with good accuracy can provide acceptable line tension prediction over the entire range of fairlead angles.

Both methods can provide tension data that can be entered in a mooring analysis together with metocean data. Such an analysis may be used to confirm actual behavior that is consistent with the design. Also, the tension measurement can be used to calculate the fatigue accumulation in the top chain. In addition, records of tension history can be used as valuable records to support a life extension application, when the owner/operator seeks to extend the service life of the mooring system.

### 12.6.3 Method 3—monitoring vessel position

A reliable and cost-effective alternative to monitor the performance of a mooring system is to observe the platform's position over time [16]. In combination with monitoring the environmental conditions on site, the measured offset and bearing from equilibrium position and the vessel heading can provide instant feedback on the mooring systems effectiveness. Position monitoring can be achieved by installing a position monitoring system onboard the vessel based on differential navigation systems [i.e., Differential Global

Positioning System (DGPS)]. The accuracy of this system can be improved significantly by adding differential corrections to the system's receiver onboard the vessel from reference stations. Often, the differential corrections are provided by satellite-based differential navigation systems.

## 12.7 Monitoring devices

There are at least three types of monitoring systems available on the market, including load cell, inclinometer, and GPS-based systems. When choosing a monitoring system, the following factors may be considered:

- *Suitability for the type of floater*—Some monitoring devices are more suitable for certain types of floaters. For example, load cells can be installed in chain stoppers on spars and semisubmersibles, but may not be easily done for a turret FPSO where its chain stoppers are located underwater.
- *Ability to measure tension*—Tension measuring is a valuable function, because it provides data for verifying mooring design. It also allows for fatigue accumulation to be calculated with accuracy.
- *Ease of use*—Display of data should be in the control room on the vessel. Vessel crews may not have the expertise to interpret data as well as mooring engineers, but a well-designed user interface can help them understand the condition of the mooring system. An example of a user interface is shown in Fig. 12.10.

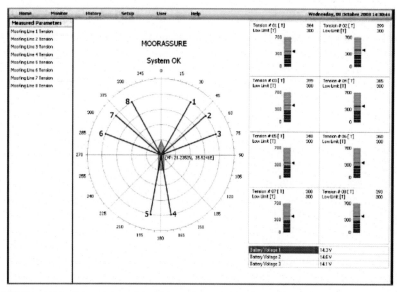

**FIGURE 12.10** Measured tension of each mooring line is displayed on a computer screen. *Courtesy: PULSE.*

- *Detection interval to allow prompt warning*—A monitoring device may or may not need frequent sampling. Less frequent sampling can be an advantage to battery life of sensors that cannot be hard-wired, but care should be taken to set the detection interval appropriately. Once a failure has occurred, it should be alarmed by the monitoring device within a reasonable time frame.
- *Reliability and track record of the device*—The device should be reliable, requiring minimal maintenance, and engineered to withstand extreme storm events. A proven design with good track record is preferred.

### 12.7.1 Load cell

Load cells are typically installed in chain stoppers on spars and semisubmersibles. They are easy to access, and thus easy to maintain, compared with the other types of devices that are placed underwater. One of the benefits of the device is that the load range required for each cell can be reduced by using multiple load cells. It allows the tension in the mooring line to be accurately measured over a larger tension range. In addition, if each load cell contains two or more sensors, the reliability of the device can be increased, which reduces the need for changing out load cells. One drawback with load cells is the uncertainty in the calibration, which has been an issue in some applications.

### 12.7.2 Inclinometer

An inclinometer measures the angle of a mooring line inclination. Many mooring systems were fitted with inclinometers that are designed to report line tensions. These systems can calculate the mooring line tension based on the angle by reference to a catenary equation. Note that the positions of the anchor and the vessel must be known in order to get the line tension converted from the angle measurement. An alarm indicating a mooring line failure would be raised when the angle went outside the predefined limits. Some inclinometer systems had poor reliability records, where some of them generated spurious readings and some failed to work [17].

The setup of an inclinometer system is arranged by fitting one inclinometer on each leg. It is mounted on either a hawse pipe or a chain (as shown in Fig. 12.11). Each inclinometer contains its own battery, memory, and acoustic emitter. The measured data is transmitted to receivers and then hard-wired to the computer display in the control room. Note that inclinometers attached to hawse pipes may not record the exact angle of the top chain, because there is a small clearance gap between chains and hawse pipes. Hence a small error may be introduced into the tension calculation.

**FIGURE 12.11**   Inclinometers installed on top chain. *Courtesy: PULSE.*

Robustness of the inclinometer systems has been a problem historically. The instrumentation components are often fully exposed to the seawater and harsh environment in which the mooring lines are located. Water ingress, corrosion, cable connections, acoustics, and battery life are some of the issues that have caused malfunctions of a system. These issues need to be carefully considered during the selection process in the engineering phase.

### 12.7.3    Global Positioning System—based system

For floating systems that do not have a tension measurement device, a device for detecting mooring failure should be considered as a minimum. Research has shown that excursion monitoring systems using GPS alone may not be sufficient to immediately detect single line failures [16]. However, the use of a DGPS to monitor vessel excursion (offset) and heading can detect a subtle change of vessel movement behavior, because a DGPS has a much higher resolution than a regular GPS. Any quick drift in the order of minutes that is unrelated to an environmental or external force could be interpreted as a potential mooring line failure. Fig. 12.12 shows a clear change in mean vessel position due to a single line break in a computer simulation [16]. Research on the utilization of DGPS for monitoring line breaks is promising. It may provide a stand-alone solution that may replace or supplement the conventional systems mentioned in the sections above. A prototype has been developed and installed, and Fig. 12.12 (right) shows one of the two antennas of the DGPS mooring monitoring system installed outside the control room of an FPSO.

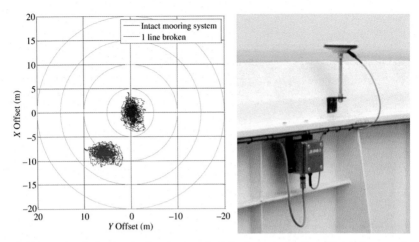

**FIGURE 12.12** (Left) Vessel offset due to a single broken mooring line. (Right) Antenna of a DGPS monitoring system outside the control room of an FPSO. *DGPS*, Differential Global Positioning System; *FPSO*, floating production storage and offloading. *Courtesy: SOFEC.*

## 12.8 Questions

1. Class Rules have an inspection framework that includes three types of in-service surveys. What are the three types?
2. With regard to conducting in-service inspections, what is the main difference between MODU moorings and permanent moorings?
3. Name at least three kinds of chain damage that an inspector should look for during an inspection.
4. An inclinometer is fitted at the top end of a 400-m chain for measuring the angle. The studless chain is 3-in in diameter, and gets deployed in water 100-m deep. Use the catenary equation in Chapter 4, Mooring design, to create a look-up table (or chart) that shows the relationship between the angle and the line tension.
5. Name two devices that are commonly used for monitoring mooring line tensions.

## References

[1] API RP 2I, Recommended Practice for In-Service Inspection of Mooring Hardware for Floating Drilling Units, first ed., American Petroleum Institute (API), 1996.
[2] API RP 2I, In-Service Inspection of Mooring Hardware for Floating Structures, third ed., American Petroleum Institute (API), 2008.
[3] DNVGL-OS-E301, Position Mooring, edition July, 2015.
[4] American Bureau of Shipping, Rules for Building and Classing Floating Production Installations, 2018.
[5] Bureau Veritas, Rule Note NR 493 DT R03 E "Classification of Mooring Systems for Permanent and Mobile Offshore Units", 2015.

[6]  J. Rosen, A. Potts, E. Fontaine, K. Ma, R. Chaplin, W. Storesund, SCORCH JIP—feedback from field recovered mooring wire ropes, in: OTC 25282, OTC Conference, May 2014.

[7]  E. Fontaine, J. Rosen, A. Potts, K. Ma, R. Melchers, SCORCH JIP—feedback on MIC and pitting corrosion from field recovered mooring chain links, in: OTC 25234, OTC Conference, May 2014.

[8]  E. Fontaine, A. Potts, K. Ma, A. Arredondo, R. Melchers, SCORCH JIP: examination and testing of severely-corroded mooring chains from West Africa, in: OTC 23012, OTC Conference, 2012.

[9]  K. Ma, R. Price, D. Villanueva, P. Monti, K. Tan, Life extension of mooring system for benchamas explorer FSO, in: Proceedings of the 19th SNAME Offshore Symposium, Society of Naval Architects and Marine Engineers, Houston, TX, February 2014.

[10]  M. Brown, A. Comley, M. Eriksen, I. Williams, P. Smedley, S. Bhattacharjee, SS: mooring system integrity: phase 2 mooring integrity JIP—summary of findings, in: Offshore Technology Conference, 2010.

[11]  K. Ma, A. Duggal, P. Smedley, D.L. Hostis, H. Shu, A historical review on integrity issues of permanent mooring systems, in: OTC 24025, OTC Conference, May 2013.

[12]  M. O'Driscoll, H. Yan, K. Ma, P. Stemmler, Replacement of corroded mooring chain on an FPSO, Ship Production Committee, SNAME Maritime Convention, 2016.

[13]  D. Witt, K. Ma, T. Lee, C. Gaylarde, S. Celikkol, Z. Makama, et al., Field studies of microbiologically influenced corrosion of mooring chains, in: OTC 27142. OTC Conference, May 2016.

[14]  K. Ma, R. Garrity, K. Longridge, H. Shu, A. Yao, T. Kwan, Improving reliability of MODU mooring systems through better design standards and practices, in: OTC 27697, OTC Conference, May 2017.

[15]  Oil & Gas UK, Mooring Integrity Guidance, Rev. F, 2008.

[16]  J. Minnebo, P. Aalberts, A. Duggal, Mooring system monitoring using DGPS, in: Proceedings of the ASME 2014 33rd International Conference on Ocean, Offshore and Arctic Engineering, OMAE 2014-24401, San Francisco, CA, 2014.

[17]  M. Brown, T.D. Hall, D.G. Marr, M. English, R.O. Snell, Floating production mooring integrity JIP-key findings, in: Offshore Technology Conference, 2005.

# Chapter 13

# Mooring reliability

## Chapter Outline

13.1 Mooring failures around the
world 256
13.2 Probability of failure for
permanent moorings 261
  13.2.1 Estimated $P_f$ for
permanent moorings 262
  13.2.2 System versus
component failures
(multiline vs
single-line breaks) 263
13.3 Failure spots for permanent
moorings 264
13.4 Probability of failure for
temporary moorings 265
  13.4.1 Estimated $P_f$ for mobile
offshore drilling unit
moorings 266
  13.4.2 Improving mobile
offshore drilling unit
mooring reliability 267
13.5 Failure spots for temporary
moorings 269

13.6 Reliability of mooring
components 270
  13.6.1 Percentage distribution
of mooring failures by
component type 270
  13.6.2 Percentage distribution
of chain failures by
cause 272
13.7 Wide variety of failure
mechanisms 273
  13.7.1 Deficient chain from
manufacturing 274
  13.7.2 Chain with severe
corrosion 274
  13.7.3 Fatigued chain due to
out-of-plane bending 275
  13.7.4 Knotted chain due to
twist 275
  13.7.5 Chain damaged from
handling 276
  13.7.6 Operation issues 276
13.8 Questions 277
References 277

Although technological advancements have enabled the migration of offshore moorings from shallow to deep water, failures of mooring lines have been occurring more frequently than would be expected when compared to the notional design expectations or other offshore structures such as fixed platforms. The industry has responded by assessing methodologies in design standards and collaborating/sharing lessons learned in open industry forums. There were 365 floating production/storage systems in operation in 2012. The number has been steadily increasing year by year. While these floating facilities have been performing well in general, their mooring systems have experienced many problems, some significant. Many of those problems

Mooring System Engineering for Offshore Structures. DOI: https://doi.org/10.1016/B978-0-12-818551-3.00013-2

are reviewed in this chapter including 26 mooring incidents that happened to floating production units (FPUs) between 2001 and 2012. The probability of system (i.e., multi-line) failure was estimated to be in the order of $10^{-3}$ per year. Several of the single-line failures included damage to the other mooring lines, thus these incidents could be counted as system failures as well. In short, the reliability record for permanent moorings has been less than satisfactory. Meanwhile, the reliability record for mobile offshore drilling unit (MODU) moorings was even worse. The probability of failure for MODUs was estimated to be in the order of $10^{-2}$ per year. The purpose of this chapter is to explain the issues and summarize the ways to mitigate them.

## 13.1 Mooring failures around the world

Mooring incidents have been occurring at a high rate. During the period between 2001 and 2012, at least eight permanent mooring systems experienced multiple-line damages, or system failure. Some of them led to vessels drifting off location. It has been found that many incidents with single-line breakage often have additional lines that sustained damage and could have also failed prematurely if their damage went undetected. Some of the incidents were of high consequence causing the vessel to drift some distance, risers to rupture, production to shut down, and even releasing a small amount of hydrocarbons. Some required substantial effort to repair or replace damaged lines. These incidents are raising concerns for owners and operators.

The eight major incidents summarized by Ma et al. [1] are listed below. A complete list including other incidents with single-line failures is summarized in Table 13.1. On a positive note, even though the incidents listed below are multiline failures, none had its mooring system broken entirely so as to allow complete vessel drifting, nor did any failure lead to a major injury, loss of life or major pollution. The information of these incidents can mostly be found in public literature:

- 2011, Banff: *5 of 10* lines parted [9,11].
- 2011, Volve: *2 of 9* lines parted, no damage to riser [12].
- 2011, Gryphon Alpha: *4 of 8* lines parted; vessel drifted a distance, *riser broken* [13].
- 2010, Jubarte: *3* lines parted between 2008 and 2010 [3].
- 2009, Nan Hai Fa Xian: *4 of 8* lines parted; vessel drifted a distance, *riser broken* [14,15].
- 2009, Hai Yang Shi You: *Entire* yoke mooring column collapsed; vessel adrift, *riser broken* [15,16].
- 2006, Liuhua (N.H.S.L.): *7 of 10* lines parted; vessel drifted a distance, *riser broken* [15,17].
- 2002, Girassol buoy: *3 (+2) of 9* lines parted, no damage to offloading lines (2 later) [18–20].

**TABLE 13.1 Failures of permanent moorings between 2001 and 2012 and their likely causes [1].**

| Year | Name (floater type) | Damaged component | Age of component | Incident | Likely causes | Water depth (ft.) |
|---|---|---|---|---|---|---|
| 2012 | Kuito (Buoy) | Chain | 13 years | 1 of 6 lines broke at suspended lower chain | Fatigue failure. Small corrosion pits at the link's crown likely allowed a crack to initiate | 1260 |
| 2012 | Petrojarl Varg (FPSO) | Chain | 6 years | 1 of 10 lines broke during heavy seas [2] | Broken chain link was exposed for OPB. High-cycle, low stress fatigue initiated on the link [2] | 280 |
| 2012 | Norne (FPSO) | Chain | 6 years | 1 line failure. With ROV inspections, line 9 was confirmed a failure [2] | Fatigue by abnormal loads or bending of the chain in 8–9 m seas. Bending could have been caused by the fairlead not rotating [2] | 1250 |
| 2012 | Haewene Brim (FPSO) | Wire rope | 8 years | 1 line failure. 3–4 additional lines with wire damage | Birdcage failure | 280 |
| 2012 | Dalia (FPSO) | Chain | 6 years | 1 of 12 lines parted in bottom (pile) chain, in a similar manner to the one in 2008 | Chain might have been knotted in the mud | 4270 |
| 2011 | Banff (FPSO) | 5 Chains | 12 years | 5 of 10 lines of the turret mooring parted. Vessel drifted 250 m off location during severe weather | | 300 |

*(Continued)*

**TABLE 13.1 (Continued)**

| Year | Name (floater type) | Damaged component | Age of component | Incident | Likely causes | Water depth (ft.) |
|---|---|---|---|---|---|---|
| 2011 | Volve—Navion Saga (FPSO) | 2 Wire ropes | 3 years | 2 of 9 lines parted at bottom end of upper wire segment at the bend stiffeners. Discovered during inspection | Ductile overload of wire rope at the termination, resulting from high local dynamic snapping loads | 270 |
| 2011 | Gryphon Alpha (FPSO) | 4 Chains | 19 years | 4 of 8 lines parted in chains in heavy storms. Vessel drifted a distance causing risers to break | 100-mph wind gusts; possible flaw in flash weld of chain link | 400 |
| 2011 | Fluminense (FPSO) | Chain | 8 years | 1 of 9 lines parted in a top chain | Likely due to an initial damage sustained in a link when flame torch was used to cut a sling before installation | 2600 |
| 2010 | Jubarte (FPSO) | 3 Chains | 2 years | 3 lines parted in lower chain segments. Between 2008 and 2010, a failure of the mooring lines Nos. 3, No. 4, and No. 5 was identified in the FPSO mooring [3] | Use of materials from different links to stud generated corrosion leading to increased stresses and failure by fatigue | 4400 |
| 2009 | Hai Yang Shi You 113 (FPSO) | Yoke column | 5 years | Yoke tower collapsed. Vessel drifted a distance causing risers to break | Strong wind; fatigue crack in the base of yoke column | 60 |
| 2009 | Nan Hai Fa Xian (FPSO) | 4 Wire ropes | 19 years | 4 of 8 lines parted in bottom end of upper wire segments in a sudden typhoon. Vessel had no time to disconnect from its BTM buoy. Vessel drifted a distance causing risers to break | Typhoon; disconnectable FPSO could not disconnect in time; overloading mooring lines; degradation of wire ropes | 380 |

| Year | Name | Component | Duration | Description | Cause | |
|---|---|---|---|---|---|---|
| 2009 | Fluminense (FPSO) | Connector | 6 years | 1 of 9 lines parted | One-line failure due to improperly installed polyester connection | 2600 |
| 2008 | Dalia (FPSO) | Chain | 2 years | 1 of 12 lines parted in bottom (pile) chain, that is, between 5 and 7 m below the mudline. Discovered during a diver operation [4]. Note another failed in a similar way later in 2012 | Chain might have been knotted in the mud | 4270 |
| 2008 | Balder (FPSO) | Chain | (9 months) | 1 of 10 lines parted in chain | Crack initiated from possible flow in chain link | 410 |
| 2008 | Blind Faith (Semi) | Connector | (0 month) | 1 of 8 lines parted | Design flaw | 6500 |
| 2007 | Tahiti (Spar) | connector | (0 month) | No lines parted. However, all anchor piles were replaced due to metallurgical issue with the shackles on them | Low fracture toughness | 4100 |
| 2007 | Kikeh (Spar) | Connector | (2 months) | 1 line parted in shackle on anchor [5,6]. Other shackles from the same batch showed low toughness | Low fracture toughness | 4400 |
| 2006 | Schiehallion (FPSO) | Chain | 8 years | 1 of 14 lines parted in chain inside hawse pipe. Later inspection discovered similar cracks in 3 other lines [7] | OPB; fretting from proof loading | 1300 |
| 2006 | Liuhua—Nan Hai Sheng Li | 7 Wire ropes | 10 years | 7 of 10 lines parted in a typhoon. Vessel drifted a distance causing risers to break | Typhoon exceeding design limit; degradation of wire ropes | 980 |
| 2006 | Varg (FPSO) | Chain | 7 years | 1 of 10 lines parted in chain [8,9] | SRB corrosion | 280 |

(Continued)

**TABLE 13.1** (Continued)

| Year | Name (floater type) | Damaged component | Age of component | Incident | Likely causes | Water depth (ft.) |
|---|---|---|---|---|---|---|
| 2005 | Kumul (Buoy) | Wire rope | 4 years | 1 of 6 lines parted in wire rope. Later inspection revealed another leg was also damaged [10] | Wire ropes in contact with seabed creating trenches from their movement | 60 |
| 2005 | Foinaven (FPSO) | Chain | 9 years | 1 of 10 lines parted. 2 other lines sustained cracks [8,9] | Corrosion fatigue on chains initiated at pitting. Rate enhanced by SRB. Stress corrosion cracking—hydrogen embrittlement | 1500 |
| 2003 | Girassol (Buoy) | Chain | 2 years | 1 of 9 lines parted in its third OPB event. Total 5 breaks in 3 events during 2002 and 2003 | Out-of-plane bending | 4599 |
| 2002 | Girassol (Buoy) | 3 Chains1 Polyester | (8 months) (10 months) | 3 of 9 lines parted, 2 OPB chain failures in hawse pipe and 1 polyester failure. Buoy drift outside of design envelope without damage to oil offloading lines. One month later, 1 more OPB. Total 4 breaks in 2 events in 2002 | Out-of-plane bending | 4600 |
| 2001 | Harding (STL buoy) | Connector | | 1 of 9 lines parted at a tri-plate-socket assembly. Socket retaining pin displaced | Poor design of face plate of the pin | 360 |

FPSO, Floating production storage and offloading; OPB, out-of-plane bending; ROV, remotely operated vehicle; SRB, sulfate-reducing bacteria; STL, submerged turret loading.

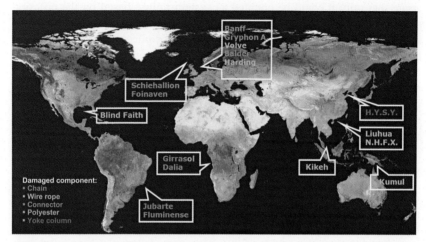

**FIGURE 13.1**  Mooring incidents in different regions around the world between 2001 and 2011 [1].

Failures of permanent moorings are distributed geographically around the world as shown in Fig. 13.1. Regions with a harsh environment, such as the North Sea, seem to experience a bigger share of the failures. In contrast, regions with a mild environment, such as the Gulf of Thailand, seem to have no major incidents. The figure is color-coded, and it shows that chain, in orange, seems to cause more problems than the other mooring components.

In addition to the failures summarized in Table 13.1, it is worth mentioning that there were two other incidents where the failures did not occur in mooring lines. One incident happened in 2009 when a new production semi was completing the final step of its mooring installation. One of the windlasses failed and caused the chain to slip out. It resulted in the dropping of one complete mooring line, including chain, polyester rope, and connectors, to the seafloor [21]. Another incident happened in early 2011 when a self-standing riser collapsed and fell to the seafloor because of a chain break. The riser was designed to be held permanently by a large air-can buoy with a tether chain. One of the chain links fractured near its flash butt weld. Analysis of the fracture indicated that the chain link had a weld repair that is not allowed by industry standards [22]. Both incidents can be arguably classified as mooring failures, especially the former. Nonetheless, these two incidents highlight that it is important to ensure the robustness in both on-vessel equipment and common mooring components.

## 13.2 Probability of failure for permanent moorings

Depending on how a failure is defined, the average failure rate per year can be expressed differently. Of the 26 incidents listed in Table 13.1, one can argue that only 8 could be counted as system failure based on the criterion of multiple-line breakage. One incident had no breakage, but the defective shackles did cause the replacement of all anchor piles. The rest are single-line breakages which could be considered as component failures. However, it should be noted that several of the single-line incidents actually had damage sustained in other lines, so more than eight of those single-line incidents could also be arguably counted as system failures. Additionally, there could be undisclosed incidents that were not documented in public literature.

### 13.2.1 Estimated $P_f$ for permanent moorings

To make a rough estimate on the failure probability ($P_f$) for permanent mooring systems, one could use a ballpark estimate of one system failure per year for the period from 2001 to 2012. There were roughly 200 permanent mooring systems around the world in 2001. They provided station-keeping to various types of floating facilities such as floating production storage and offloading (FPSO); floating, storage, and offloading (FSO); Semi; spar; and buoy. There were a total of 286 floating production systems around the world in 2010, according to the counts made by *Offshore Magazine* [23]. That count does not include offloading buoys or storage vessels, for example, FSO. In 2012 there were 365 floating production/storage systems in service according to a report published by International Maritime Associates [24], not counting offloading buoys. To conduct a rough estimate, it should be fair to assume that the total number of permanent mooring systems was increasing from 200 to about 400 during the period from 2001 to 2012. With around nine major mooring incidents out of the averaging 300 permanent mooring systems in a 12-year period, the annual probability of failure could be estimated as the following:

$$P_f \approx 2.5 \times 10^{-3}$$

**Note 1:** $P_f$ is an estimate of the achieved annual probability of failure during the 12-year period from 2001 to 2012 among the approximately 200−400 permanent mooring systems.

**Note 2:** Failure is defined as any incident involving (A) breakage of two or more lines, or (B) riser damage. With this definition, an incident would not be counted as a failure if there is only a single-line break.

**Note 3:** Permanent mooring systems are those used on long-term floating facilities such as FPSO, FSO, FPU, spar, production semi, and offloading buoys. Chains and ropes used on self-standing risers, mid-water arches, or subsea equipment are not considered.

It is believed the probability of failure is too high and the mooring industry has room for improvement. Note that the achieved $P_f$ would be higher (worse) if all 21 incidents in Table 13.1 were counted.

## 13.2.2  System versus component failures (multiline vs single-line breaks)

Regarding the definition of mooring failure, there are various versions and they can be confusing. Two-line breakage should be typically considered as a system failure if a single-line damage criterion was incorporated in the design. However, if a design was based on a two-line damage criterion, a breakage of two lines was still within the design and should not be considered as a system failure. The key to distinguish whether a mooring system had a failure or simply sustained some damage is to check if it can still protect subsea equipment from any ill effect until the next scheduled inspection. Using this concept, a mooring system failure may be divided into the following two types:

1. *Multiple-line breakage causing riser damages*

    Table 13.1 shows that four incidents had such system failures, that is, Gryphon Alpha, Nan Hai Fa Xian, Liuhua, and Hai Yang Shi You. They were mostly related to extreme environmental conditions, such as typhoons, winter storms, or strong winds. Fig. 13.1 shows that two of the four are located in the typhoon-prone South China Sea, while Gryphon Alpha is in the harsh environment of the North Sea. In the four cases, vessels all drifted a sufficient distance to pull their risers to rupture. Although there has been no record of any casualties, the consequence is high due to the prolonged production shutdown for repair.

2. *Multiple-line breakage without riser damages*

    There are at least four incidents with multiple-line breaks that fortunately incurred no damage to risers. These include Banff, Volve, Girassol, and Jubarte.

Since mooring systems are usually designed with one-line break specifically considered, single-line breakage can be considered as a component failure in the system. Often, engineers refer to these as "damage" rather than failures. The primary function of the mooring system is to keep the vessel on station so that risers, umbilicals, and subsea equipment are protected from damage. Following industry codes or Class Rules, a mooring system with one line parted should still be capable of serving its station-keeping function at a reduced safety factor. Such a condition should not immediately threaten the integrity of risers. In Table 13.1, there are many cases of single-line breakages, and all managed to keep their risers intact after parting one line.

## 13.3 Failure spots for permanent moorings

For permanent mooring systems, failures can occur at any point along a mooring line, as shown in Fig. 13.2. However, in the majority of instances, failures occurred at an interface or discontinuity. Such an interface can be at the fairlead between the line and the vessel, at connectors between types of lines, or at connections for spring buoys, clump weights, tri-plates, etc. Failures can also occur where the mooring line dynamically touches the sea-bed (thrash zone) or where the line descends into the mud to link to an anchor or pile.

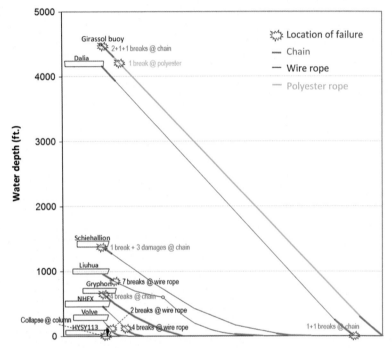

**FIGURE 13.2** Location of breaks along mooring profiles [1].

During design, the mooring line is modeled as a simple tension-only element with section properties reflecting the components along the line. Consequently, compression, bending, and torsion have been ignored, but each has been found to be either a primary or contributory cause of line failure.

It is interesting to note that several of the incidents in Table 13.1 had the same damage mechanism occurred on multiple lines, and at the same locations along the line profile. In other words, the presence of a weak point often applies to all lines, increasing the likelihood of multiple-line failure.

For example, the four breaks in the Nan Hai Fa Xian incident all occurred at the wire rope terminations where the wire rope was connected to the ground chain. This demonstrates the importance of a robust design that is free of weak points.

While some of these failures may be prevented through more robust inspection and monitoring, one could argue that many of them were novel phenomena in nature. For instance, the out-of-plane bending (OPB) on Girassol's mooring chains was a newly discovered phenomenon in 2002 [18−20]. Conventional remotely operated vehicle (ROV) surveys would have been unable to detect any fatigued chain links, as these were hidden inside hawse pipes at the top end of a mooring line.

Another example is the broken shackle on one of Kikeh's anchors that also came as a surprise to the industry. The issue with low fracture toughness was not something that could be detected by any kind of in-service inspection, not to mention the fact that the defective shackle was totally buried in the mud.

Note that some failures occurred in corroded wire ropes, and the corrosion was very visible in those unsheathed wire ropes. In one case, the degradation had been detected by ROV inspection, but had not been replaced in time before they broke in a storm.

## 13.4    Probability of failure for temporary moorings

MODU is just one type of floating facility that uses temporary moorings, also referred to as short-term or mobile moorings. Other types include workover barges, construction barges, floatels, tender-assisted-drilling vessels next to another platform, etc. They all have planned field service time ranging from a few weeks or a few months to typically no more than 12 months. While this section focuses on MODU, most of the discussion points are applicable to all types of temporary moorings.

MODU moorings are designed to a much lower return period, typically 5−10 versus 100 years for permanent moorings and therefore have much weaker mooring components. Unlike the permanent moorings, which almost seldom fail in overloading, MODU moorings can fail in overloading in severe tropical cyclones, that is, hurricanes and typhoons. MODU mooring failure data for "worldwide" operations have not been systematically collected and studied. Many MODU mooring failures are not reported, and therefore accurate data are not available. However, based on available information from the Gulf of Mexico (GOM) and offshore Australia, the MODU mooring failure rate can be considered approximately one order of magnitude higher than permanent mooring failure rate.

### 13.4.1 Estimated $P_f$ for mobile offshore drilling unit moorings

MODUs normally stay at one location for a shorter term, compared to tens of years for permanent moorings on FPUs. While the exposure time to the environment is relatively shorter, mobile moorings have been seen to iron-ically experience a sizable number of failures. After being hit by hurri-canes in GOM between 2004 and 2008, many mobile moorings failed and caused those MODU rigs to drift [26,27]. A paper by Stiff [28] summa-rized that "The annual probability of complete mooring failure of a MODU operating in the GOM (1980−2008), defined as the unit moved more than one mile from its original location, was *calculated to be* $3.8 \times 10^{-2}$." In contrast, the previous section discusses that *The annual probability of failure of a permanent installation is approximately* $2.5 \times 10^{-3}$. That is roughly an order of magnitude difference between MODU and permanent moorings. While one could argue that the defini-tion of failure is a bit different and thus a direct comparison may not be fair, the high failure rate of MODU moorings was still disconcerting to the mooring experts in the industry.

In reaction to the drifting MODUs in GOM, a Joint Industry Project (JIP)—"Gulf of Mexico MODU Mooring"—was formed to assess the reli-ability of MODU moorings. Ultimately, the lessons learned were rolled into API RP-2SK in 2008. Appendix K "MODU for GOM Hurricane Season" was developed in 2SK to augment the requirements on metocean return periods for MODU mooring in hurricane seasons. Since then, the majority of MODUs in the GOM have been upgraded. For example, some upgrades were accomplished by adding one preset line to each corner of the MODUs. Those extra mooring lines are sometimes referred to as "storm" legs. The addition of Appendix K into the API mooring codes was an improvement in coping with extreme environmental conditions. To a large degree, it was an effective upgrade in codes, and the result has proved favorable.

Failure of mobile moorings still occurs unfortunately. A few incidents happened in cyclone areas offshore Australia, for example. An information paper [29] released by the Australian regulatory agency indicated that there were four incidents between 2004 and 2015 where the impact of cyclone activity resulted in the loss of position of a moored MODU in Australian waters. Using a very rough estimate of averaging eight rigs per year operat-ing in the region, the annual probability of failure ($P_f$) is calculated to be *approximately* $4.2 \times 10^{-2}$. The $P_f$ number matches the $3.8 \times 10^{-2}$ calculated for GOM quite well, while it should be noted that the four failures in Australia includes anchor drags of any distance.

If failure is strictly defined as drifting more than one mile, the number of failures arguably drops from four to one, which corresponds to $P_f$ of *approxi-mately* $1.0 \times 10^{-2}$ [30]. Nonetheless, the failure probability is clearly still too high. It is a potential gap that needs to be closed, and there is clearly room for improvement. Based on the numbers discussed above, the annual probability of failure for MODU moorings may be estimated as the following:

$$P_f \approx 1.0 \times 10^{-2}$$

**Note 1:** $P_f$ is a rough estimate of the achieved annual probability of failure based on data available in the literature for MODU moorings in GOM and Australian waters.

**Note 2:** Failure is defined as any incident where the MODU moved more than one mile from its original location.

Although industry standards published by API, ISO, and Class provide criteria for mooring design, they may not be comprehensively effective in preventing mooring line failures from happening. In areas of mild environment, single-line failures can be seen by operations occasionally. Sometimes, operators get line failures so often to the point that they start to believe that line failure is a part of the normal operation and is somewhat expected. When these incidents happen on mobile moorings, they do not receive as much attention as those on permanent moorings. On the other hand, mobile moorings are watched more closely in areas of tropical cyclones, and may receive more attention. Multiple-line failures can occur there due to extreme environmental forces, and can lead to a vessel drifting.

## 13.4.2  Improving mobile offshore drilling unit mooring reliability

Developing design criteria for mobile moorings is not a simple task. It is as sophisticated as developing criteria for permanent moorings. Probability and the consequence of a failure need to be taken into account jointly. Target failure probability, or target reliability, needs to be assessed to set the required minimum return period in codes. It is always controversial how high a target should be set. The failure consequence of a mobile mooring at close proximity can be as high as that of the neighboring infrastructure, even though its exposure time to the storm condition is notably shorter. When hit by extreme storms such as a tropical cyclone, MODU moorings are often more vulnerable compared to permanent moorings, and can experience multiple-line breaks resulting in severe consequences. An incident can cause rig downtime for weeks or even months. More importantly, there is the possibility that vessels could drag remaining mooring lines and anchors resulting in damage to infrastructure nearby.

Improving reliability of MODU moorings may be achieved on two fronts, that is, improving design standards and applying more rigorous operation practices. On the design front, there appears to be a lack of clear guidance in industry codes on designing a mobile mooring system to a proper return period. On the operation practice front, MODU moorings often do not receive enough attention in system design, deployment, inspection, and equipment maintenance. Obviously, MODU operators need to better understand typical failure mechanisms, and stay cautious in how to prevent those from happening.

MODU mooring has experienced a relatively high annual probability of failure. There is room for improvement, and the reliability of MODU moorings is in need of a promotion. A higher target reliability needs to be aimed at, as shown schematically in Fig. 13.3. It can be achieved by lowering the probability of failure to below $10^{-2}$ and by reducing the consequence of failure with implementation of effective mitigations.

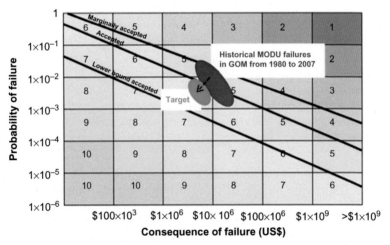

**FIGURE 13.3** Target reliability may be improved for MODU mooring [30]. *MODU*, mobile offshore drilling unit.

Although significant efforts have been devoted to improving the reliability of MODU moorings, there are limitations in these efforts. Design return period could not be raised easily, because it would potentially impact the availability and mobility of the MODUs for drilling operations. Better practice in manufacturing, inspection, repair, replacement, and installation relies on personal behaviors, and therefore may be difficult to achieve.

## 13.5 Failure spots for temporary moorings

In a permanent mooring system, failure can potentially initiate from any individual component. For mobile moorings, wire ropes seem to be one of the most problematic components, while chain and connectors contribute to a much smaller share of the problems. A study in 2009 analyzed MODU mooring failures in hurricanes Katrina and Rita [26]; it concluded "The most striking factor is the number of mooring lines that fail at the fairlead. *Over 80% of the line failures listed occurred at, or close to, the fairlead.*"

Six-strand or eight-strand wire ropes are widely used as MODU mooring lines in deep water to keep the vessels anchored on station, as they are lighter than chains and easy to handle and deploy. Unlike chain that is made of solid steel, wire ropes are made of a bundle of small wires and by their physical makeup are prone to damage and require considerable attention to assure integrity for safe operations. Based on experience and learnings from a couple of JIPs [31−33], it is found that wire ropes have at least two weak points as shown in Fig. 13.4.

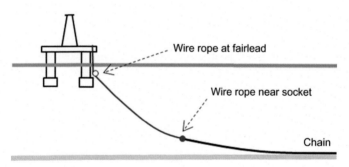

**FIGURE 13.4**  Critical spots in a mooring wire rope [30].

The first weak point is the part passing through the sheave on a fairlead. A wire rope sees the highest tension near the top of the line, because of its own weight. Moreover, a wire rope on a sheave gets additional bending and compression stresses [34]. Most wire rope failures occur at this spot. An example of such a failure is shown in Fig. 13.5. Moreover, additional attention needs to be paid to undersized sheaves. If the sheave is undersized, the bending fatigue can accelerate the degradation process significantly. *D/d* ratio (fairlead sheave diameter over wire rope diameter) should not be lower than minimum values recommended by standards [25,35], such as 16 for mobile moorings and 40 for permanent moorings. Damaged wire rope due to low *D/d* ratio can often be seen on some of the older work barges.

**FIGURE 13.5** Wire rope often fails prematurely at fairlead [30].

The second weak point is at socket terminations. Broken wires can often be found at this location. Wires near the socket termination experience a lot of cyclic bending and torsional loads such that the wires at the outer layer eventually fatigue due to localized stress concentrations. If a socket termination is located near the touchdown zone in a mooring profile, it can suffer from repeated beatings on the seabed and could break quickly. Such a mooring configuration should be avoided and socket terminations should be inspected after each deployment.

## 13.6 Reliability of mooring components

As an engineer working on a mooring project, it is beneficial to know what components tend to have more problems. Such a topic has been discussed in several technical papers [1,30,31,33,36].

### 13.6.1 Percentage distribution of mooring failures by component type

The three components that cause the most incidents in permanent moorings are chains, connectors (including shackle, H-link, tri-plate), and wire ropes [1]. Fig. 13.6 shows the failure percentage distribution by component type from 2001 to 2012. It can be seen that *chain failures contribute to 54% of the incidents*. In other words, roughly half of the mooring incidents happened in chain. This high percentage may be explained by the sheer number of chain links installed in the field around the world for both shallow and deep waters. It is also observed that the chain manufacturing process is of high complexity which may have contributed to the high percentage.

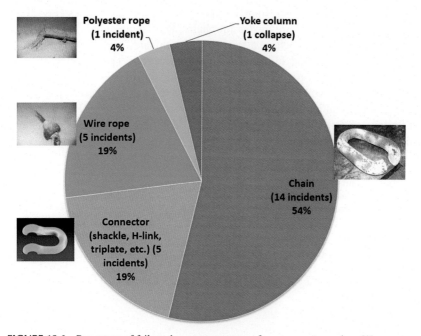

**FIGURE 13.6**   Percentage of failures by component types for permanent moorings [1].

Following chain, connectors and wire rope also contribute a large share of the mooring incidents over the same time period (2001−12). Designs of connectors have been improving based on lessons learned, and thus their failure rates are reducing steadily over time. Wire ropes for permanent moorings can be originally sheathed (i.e., a protective outer coating) or unsheathed (i.e., bare, unprotected wire). The majority of the failed wire ropes were unsheathed. However, wire ropes are almost always sheathed nowadays on permanent moorings, and therefore wire rope failures due to corrosion should be alleviated over time.

An interesting finding is that polyester rope constitutes a surprisingly small percentage of the failures. At one time, using polyester in moorings was considered as a new and unproven technology. Time has proved that polyester ropes are very reliable and therefore are gaining a wider application. It has become the most favored component for deepwater moorings due to its light weight and good reliability. Note however there have been a few incidents where polyester ropes were accidentally damaged and cut by umbilical tether wires from an ROV's tether management system or work wires from a boat.

### 13.6.2 Percentage distribution of chain failures by cause

Since chain is the component type that contributes to most of the mooring failures, it would be interesting to look further into the causes. Fig. 13.7 shows the prevalent failure modes for chain [36]. The pie chart is based on a comprehensive survey, where a total of 10 companies contributed, reporting a total of 61 individual failures on 43 specific units. The reported failures were associated with permanent mooring systems of FPSOs, FSOs, semisubmersibles, Spars, and CALM buoys. A failure event is defined as any situation including single-line failure or multiple-line failure. For Fig. 13.7, the definition of failure event is expanded to include preemptive replacement and reported degradation.

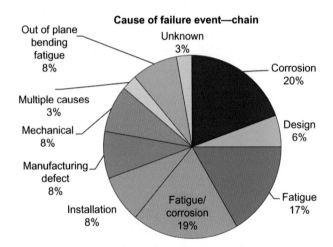

**FIGURE 13.7** Cause of failure and preemptive events for chain [36].

While the chart may appear complicated with three slices related to fatigue and two slices to corrosion, it can be reconciled that *fatigue and corrosion are the top two failure modes*. There are three slices related to fatigue, that is, 17% tension−tension fatigue, 8% OPB fatigue, and 19% fatigue/corrosion. There are two slices connected to corrosion, that is, 20% corrosion, and 19% fatigue/corrosion. Note that the slice labeled "fatigue/corrosion" counts those fatigue failures that were initiated by corrosion grooves/pits and accelerated by corrosion. Together, the combination of fatigue and corrosion contributes a total of 64% of the failure events. Failure mechanisms related to fatigue and corrosion are further discussed in the next section.

To improve mooring reliability, these two issues clearly deserve more research work in the future.

## 13.7    Wide variety of failure mechanisms

A surprising finding from the paper by Ma et al. [1] is that there is a wide variety of failure mechanisms among the past incidents that happened to permanent moorings. A common misunderstanding is that mooring lines failed because they were overloaded by extreme weather conditions. In other words, a fierce environmental event hit the mooring system, and some of the lines broke simply due to overload (overtension) because design conditions were exceeded. Surprisingly, it was found that most of the failures were not due to weather overload, but were caused by many other failure mechanisms. These failure mechanisms included OPB fatigue, pitting corrosion, flawed flash welds, unauthorized chain repair, chain hockling (knotting) due to twist, low fracture toughness, and many others. Based on the data in the paper, Fig. 13.8 was created to show the wide variety of failure mechanisms for three component types, that is, chain, wire rope, and connector.

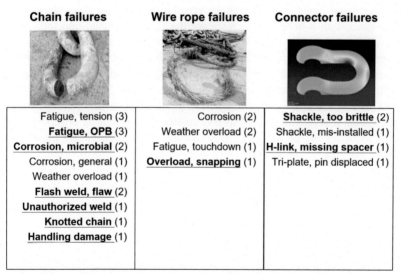

| Chain failures | Wire rope failures | Connector failures |
|---|---|---|
| Fatigue, tension (3) | Corrosion (2) | **Shackle, too brittle** (2) |
| **Fatigue, OPB** (3) | Weather overload (2) | Shackle, mis-installed (1) |
| **Corrosion, microbial** (2) | Fatigue, touchdown (1) | **H-link, missing spacer** (1) |
| Corrosion, general (1) | **Overload, snapping** (1) | Tri-plate, pin displaced (1) |
| Weather overload (1) | | |
| **Flash weld, flaw** (2) | | |
| **Unauthorized weld** (1) | | |
| **Knotted chain** (1) | | |
| **Handling damage** (1) | | |

\* **Bold underline** indicates any "novel" mechanism that is not addressed in design codes at the time of failure.

**FIGURE 13.8**  Variety of failure mechanisms for chain, wire rope, and connector with number of incidents happened during the period from 2001 to 2011.

In addition to the wide variety of failure mechanisms, the table also shows that half of the failure mechanisms are novel. They were unexpected failures due to a variety of reasons, such as unknown fatigue mode, new material issue, unproven designs of components, and corrosive seawater environment. Note that novel failure mechanisms are particularly troubling, because they may not be easily caught by any existing quality assurance/quality control procedures during manufacturing or in-service inspection practices. Also, most of the mentioned failure modes cannot be easily detected through a general visual inspection by a diver or ROV. Corrosion may be an exception that can often be detected through inspection.

Due to the novelty of some failure mechanisms, they are not addressed in the current codes and standards. With the various failure mechanisms, mooring engineers are faced with the emergence of unfamiliar problems not originally foreseen in the design of the system. Some of those may be partly due to the increasing size and grade of mooring components used for larger floating structures. Some may be due to higher pretensions used to maintain station-keeping requirements in deeper water. The following sections give some high-profiled examples of these novel failure mechanisms.

### 13.7.1 Deficient chain from manufacturing

A number of mooring lines failed due to chain deficiencies that were introduced to the surface of chain when improper weld-repairs were done by the manufacturer to patch manufacturing defects. Note that chain with manufacturing defects should not be repaired but should be scrapped. The other areas with a history of defects are the flash butt weld and loose studs on stud link chain. As chain size and grade increases, it becomes harder for the manufacturer to produce or inspect flash butt welds, and any embedded defects or lack of fusion can lead to premature fracture or fatigue failure. Manufacturers are asked to produce larger components and use higher grade steels, which sometimes compromise the ability to maintain good quality. In these cases, a higher degree of quality assurance and testing is required and needs to be agreed between the manufacturer and the purchaser prior to the start of manufacturing. Studs in stud link chain have a historical issue around fixity and fusion (i.e., loose studs). Where stud link chain is used for long-term moorings, corrosion can also lead to loose studs, and the fatigue life of stud link chain becomes shortened.

### 13.7.2 Chain with severe corrosion

It should be highlighted that corrosion has been the main reason for several preemptive replacements of mooring systems. Both general and pitting corrosion can be very damaging to top chain in certain regions of the world. Chain in the splash zone and upper water column can suffer a corrosion rate of 1.0 mm/year or higher, as depicted in Fig. 13.9. The observed corrosion rates were a surprise to mooring designers who originally applied the corrosion allowance of 0.4 mm/year required by most industry standards. Additionally, microbiologically influenced corrosion can create large pits or patches in submerged chains that are situated in the water column or on the seabed [9,38].

**FIGURE 13.9** Examples of severe corrosion on chain in splash zone [37].

### 13.7.3 Fatigued chain due to out-of-plane bending

OPB fatigue of mooring chains was identified as a new failure mechanism after the failure of several mooring chains of Girassol's deepwater offloading buoy in 2002. The discovery of this new failure mechanism resulted in a JIP that provided valuable insights as well as a design methodology. It is an engineering fundamental that one cannot bend a length of chain because links can rotate freely around adjacent links. However, it has been demonstrated that the chain can act like a beam member with lateral (pitching) motion taken as bending at the top and second links when it is under a high tension. The local bending in the top link has been confirmed to be the root cause of at least three mooring failures [39]. Note that the mechanism between the top and second links is actually quite complex, involving three distinct phases, locking, sticking-sliding, and sliding. Refer to Chapter 6, Fatigue analysis, for more details.

### 13.7.4 Knotted chain due to twist

Limiting the amount of twist in chain or rope is recommended by manufacturers. This is important during the installation phase, when very long lengths of mooring line are lowered off an anchor handler or construction vessel and connected to prelaid chain sections attached to an anchor pile. The installation contractors minimize twist in the chain through visual inspection. On the Dalia FPSO, twists led to two failures in the pigtail (forerunner) chains that run from the anchor pile to a subsea connector. A hockle (knot) is believed to have formed in the chain which ordinarily would pull out and straighten when load is applied. However, being within the seabed, the chain may have had sufficient soil resistance to prevent the straightening. As a result, the link endured tensile loads in an abnormal orientation, and failed in fatigue. Improved installation practice may have prevented this novel failure.

### 13.7.5 Chain damaged from handling

There have been examples where mistreatment or poor handling of mooring lines during transportation and installation led to failure. In one case, it was suspected that uncontrolled welding heat into a chain link introduced high local residual stresses that led to later chain fracture. Chain tends to get handled in a rough manner on the anchor handling vessel as it has the appearance of sturdiness. However, mistreatment can surprisingly cause local damage in a chain link and impact its integrity. Fiber rope is more delicate and is therefore generally handled more carefully.

### 13.7.6 Operation issues

During field operation, various types of nontechnical problems can arise in addition to those technical issues already mentioned above. One example is for a vessel with a disconnectable mooring system. The mooring system is not designed for an extreme storm, and relies on human judgment to disconnect when a storm becomes a threat. It has happened that an offshore installation manager (OIM) failed to disconnect a disconnectable mooring system before a sudden typhoon arose and hit the vessel. It needs to be understood that the OIM is always under pressure to avoid any kind of production shutdown including those due to mooring disconnects. Due to such a controversy, designers debate among themselves whether a disconnectable mooring system is a trustworthy solution. Another example of an operational issue affecting moorings is the overreliance on active heading-control thruster systems to prevent the vessel from turning beam on during storms. This latter example also requires human intervention to avoid the adverse vessel orientation. Compared to passive weathervaning turret moorings, active heading control may create too much complexity that is best avoided if possible.

In terms of operational issues that may have an adverse impact on the integrity of a mooring system, two approaches for integrity improvement may be considered. The first approach is to improve operational practices and training. Alternatively, the mooring designers can strive to minimize or eliminate the need for operator intervention.

## 13.8    Questions

1. How would you define failure for permanent mooring systems? How would you define that for temporary mooring systems?
2. Where are the typical failure spots in permanent mooring systems? Where are those in temporary mooring systems?
3. You are appointed to be the chair of an industry committee developing mooring codes and standards. Draw a chart that can be presented in a meeting to inspire committee members to improve target reliability.
4. Which mooring component caused more failures in permanent mooring systems based on the data presented in this chapter? In mobile (temporary) moorings, which component may be more problematic than others?
5. There are a wide variety of failure mechanisms. Describe at least three of those which are most intriguing to you.

## References

[1] K. Ma, A. Duggal, P. Smedley, D. L'Hostis, H. Shu, A historical review on integrity issues of permanent mooring systems, in: OTC 24025, OTC Conference, May 2013.

[2] A. Kvitrud, Anchor line failures—Norwegian continental shelf—2010−2014, Report 992081, 2014.

[3] L. Largura, L. Piana, P. Craidy, Evaluation of premature failure of links in the docking system of a FPSO, in: OMAE 49350, OMAE Conference, Rotterdam, June 19−24, 2011.

[4] TOTAL, Presentation at MCE Deep Water London, 2011.

[5] A. Cottrill, Mooring system integrity a hot button question—series of failures sparks industry investigation into materials and current manufacturing processes. <https://www.upstreamonline.com/>, 2008.

[6] MMS (BSEE), Catastrophic Failures in Mooring Systems Possibly Put Floating Structures at Risk, Safety Alert No. 259, United States Department of the Interior, January 2008.

[7] P. Smedley, Schiehallion mooring chain failure, in: Presentation at OPB JIP Meeting, 2009.

[8] Teekay Petrojarl Production, Experience with bacterial corrosion on chain, in: Ramnas Technical Seminar, 2012.

[9] Teekay Petrojarl Production, Sulphate Reducing Bacteria—Erfaring Med SRB Angrep Pa Kjetting, Tekna, Trondheim, 2012.

[10] C.R. Chaplin, A.E. Potts, A. Curtis, Degradation of wire rope mooring lines in SE Asian waters, in: Offshore Asia Conference, Kuala Lumpur, 2008.

[11] Upstream Online, Drifting Petrojarl Banff stable, December 13, 2011.

[12] S. Moxnes, Multiple Steel Wire Rope Failures on Volve FSU Mooring, Synergy No. 1231190, Safety Alert, Issued by Statoil, 2011.

[13] Maersk, Gryphon Alpha Loss of Heading, Mooring System Failure and Subsequent Loss of Position, Safety Alert, Issued by Maersk, 2011.

[14] CNOOC, No Casualties as Typhoon Koppu Blasts Huizhou Oilfields, Press Release by CNOOC, 2009.

[15] J. Wang, To build a reliability SPM, in: Presented at Second Annual Summit—Excellence in FPSO Design, Construction and Operations, CNOOC Energy Technology & Service Co., Tianjin, China, May 5, 2012.

[16] CNOOC, No Injuries No Oil Spills as Strong Wind Hit HaiYangShiYou 113, Press Release by CNOOC, 2009.

[17] A. Wang, R. Pingsheng, Z. Shaohua, Recovery and re-hook-up of Liu Hua 11-1 FPSO mooring system, in: Proceedings of the Offshore Technology Conference, OTC 19922, 2009.

[18] P. Jean, K. Goessens, D. L'Hostis, Failure of Chains by Bending on Deepwater Mooring Systems, OTC 17238, 2005.

[19] C. Melis, P. Jean, P. Vargas, Out-of-plane bending testing of chain links, in: OMAE 67353, OMAE Conference, Halkidiki, Greece, 2005.

[20] P. Vargas, P. Jean, FEA of out-of-plane fatigue mechanism of chain links, in: OMAE 67354, OMAE Conference, Halkidiki, Greece, 2005.

[21] BSEE (MMS), Accident Investigation Report—15 May 2009, United States Department of the Interior, Minerals Management Service, 2010.

[22] BSEE (MMS), Catastrophic Failures in Mooring Systems Possibly Put Floating Structures at Risk, Safety Alert No. 296, United States Department of the Interior, May 2011.

[23] Offshore Magazine, 2010 worldwide survey of floating production, storage and offloading (FPSO) units, Offshore Mag. Mustang Eng. (2010).

[24] IMA, Floating Production Systems—Assessment of the Outlook for FPSOs, Semis, TLPs, Spars, FLNGs, FSRUs and FSOs, International Maritime Associates, Inc., Washington, DC, 2012.

[25] DnV, Position Mooring, Offshore Standard, DNV-OS-E301, Det Norske Veritas, 2001.

[26] M. Sharples, J. Stiff, Metocean return period required for mooring during cyclone season, in: Offshore Technology Conference, OTC 2014, 2009.

[27] J. Stiff, MODU risk—MODU mooring comparative risk assessment, in: Offshore Technology Conference, OTC 20143, 2009.

[28] J. Stiff, How reliable are reliability calculations—illustrated with stationkeeping examples, in: Proceedings of the Offshore Structural Reliability Conference, Houston, TX, September 16—18, 2014.

[29] NOPSEMA, MODU Mooring Systems in Cyclonic Conditions, Information Paper, N06000-IP1631, December 17, 2015.

[30] K. Ma, R. Garrity, K. Longridge, H. Shu, A. Yao, T. Kwan, Improving reliability of MODU mooring systems through better design standards and practices, in: Offshore Technology Conference, OTC 27697, May 2017.

[31] M. Brown, T. Hall, D. Marr, M. English, R. Snell, Floating production mooring integrity JIP— key findings, in: Proceedings of the Offshore Technology Conference, OTC 17499, 2005.

[32] ABS Consulting, Gulf of Mexico MODU Mooring JIP, Managed by ABS Consulting, 2005.

[33] M. Brown, A. Comley, M. Eriksen, I. Williams, P. Smedley, S. Bhattacharjee, Phase 2 mooring integrity JIP—summary of findings, in: Proceedings of the Offshore Technology Conference, OTC 20613, 2010.

[34] A. Potts, C. Chaplin, N. Tantrum, Factors influencing the endurance of steel wire ropes for mooring offshore structures, in: Offshore Technology Conference, OTC 5718, May 1988.

[35] API RP-2SK, Design and Analysis of Stationkeeping Systems for Floating Structures, third ed., 2005.

[36] E. Fontaine, A. Kilner, C. Carra, D. Washington, K. Ma, A. Phadke, et al., Industry survey of past mooring failures, pre-emptive replacements and reported degradations for mooring systems of floating production units, in: OTC Conference, OTC 25273, May 2014.

[37] H. Shu, A. Yao, K. Ma, W. Ma, J. Miller, API RP 2SK 4th edition—an updated station-keeping standard for the global offshore environment, in: OTC 29024, OTC Conference, May 2018.

[38] E. Fontaine, A.E. Potts, K. Ma, A. Arredondo, R.E. Melchers, SCORCH JIP: examination and testing of severely-corroded mooring chains from West Africa, in: Proceedings of the Offshore Technology Conference, OTC 23012, May 2012.

[39] A. Izadparast, C. Heyl, K. Ma, P. Vargas, J. Zou, Guidance for assessing out-of-plane bending fatigue on chain used in permanent mooring systems, in: Proceedings of the 23rd Offshore Symposium, Society of Naval Architects and Marine Engineers (SNAME), Houston, TX, 2018.

# Chapter 14

# Integrity management

## Chapter Outline

**14.1 Mooring integrity management** 282
  14.1.1 Managing mooring performance 282
  14.1.2 Assessing hazards and performing risk assessment 283
**14.2 Incident response** 284
  14.2.1 Define response actions 285
  14.2.2 Include a sparing plan 286
  14.2.3 Predefine installation procedures and contracting plan 287
  14.2.4 Include procedures for readiness check of equipment 287
**14.3 Life extension** 287

  14.3.1 Life extension for a floating facility and its mooring system 288
  14.3.2 Fitness assessment of mooring component 289
**14.4 Ways to improve mooring integrity** 291
  14.4.1 Perform rigorous inspection and maintenance 291
  14.4.2 Equip with monitoring system 293
  14.4.3 Share lessons learned 294
  14.4.4 Improve codes and standards 294
**14.5 Questions** 295
**References** 296

This chapter discusses the integrity management of mooring systems connected to a "permanent" floating system used for the drilling, development, production, or storage of hydrocarbons. It is a process referred to as mooring integrity management or MIM. Its scope covers the entire mooring system from the anchor to on-vessel equipment. Specific guidance is provided in technical areas including monitoring, repair (that covers incident rapid response), life extension, and ways to improve mooring integrity. Note that temporary mooring systems can be inspected and maintained relatively more easily than permanent systems. Therefore, this chapter focuses on permanent mooring systems. The same principles can still apply to temporary mooring systems. MIM, along with RIM (Riser Integrity Management), can be considered as a part of the Floating System Integrity Management. However, MIM has its very unique features and issues, and often is handled independently from RIM.

Mooring System Engineering for Offshore Structures. DOI: https://doi.org/10.1016/B978-0-12-818551-3.00014-4

## 14.1 Mooring integrity management

MIM is the process of ensuring a mooring system's fitness-for-service over its entire life from engineering (design) to life extension. It is a process for managing the effects of deterioration, changes in loading and accidental overload, and responses to a line failure [1−3]. The objectives of a MIM process include detection of possible degradation or failure of a mooring component at a sufficiently early stage to allow for remedial action. The integrity management process should also provide a record of inspection, maintenance, and service data that will be required when considering future life extension.

The key components of MIM are illustrated in Fig. 14.1, where the MIM process starts as early as the engineering phase and continues all the way to a potential life extension. Note that inspection and monitoring procedures are covered separately in Chapter 12, Inspection and monitoring.

Mooring Integrity Management

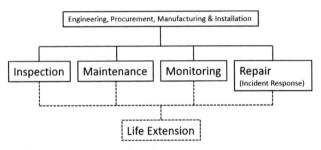

FIGURE 14.1 Components of mooring integrity management.

The integrity management process provides the opportunity for owners/ operators and their engineers to adopt risk-based principles for developing strategies that take into account the current condition of the mooring system, the likelihood of damage or degradation of a mooring line, and the potential consequences [1,2]. A risk-based approach recognizes that moorings with higher risks can warrant more frequent and more focused inspection than moorings with lower risks. During the development of an inspection strategy, the mooring risk category can be used for setting survey intervals and work scopes. The inspection work scope should take into account the latest lessons learned from all operators in the industry.

### 14.1.1 Managing mooring performance

To manage mooring integrity effectively, performance parameters need to be in place for the mooring system. A method of providing assurance for each of the mooring system's, or component's, functionalities and their criteria are necessary. These should be specific and measurable. The following three items can be considered for inclusion in the performance parameters [4].

*Vessel offset*—The purpose of a mooring system is to keep the vessel on station so that risers and umbilicals are protected from getting damaged. Vessel offset is one of the most important performance parameters for mooring. This parameter has allowable limits predefined for intact and damaged conditions. Allowable vessel offsets can be expressed as a traffic light arrangement (green, yellow, and red concentric circles) linked to actions to be taken as the offset increases. Assurance activities for offset are typically provided by a global positioning system and a display in the control room. The actual offsets for at least one period in a known metocean condition should be compared to the original design to verify whether the mooring system behaves as designed.

*Remaining strength*—The strength performance of a mooring line is indicated by a safety factor, which is typically defined as the minimum break strength divided by peak tensile load, which has been identified in the worst load case in the design. A process should be developed for assessment of the effect on remaining strength of a mooring component when any wear, corrosion, or other anomaly is identified. This allows the strength safety factor to be updated and compared with the requirements in codes and standards. Note that initial design verification activities for component strengths typically take place during manufacturing phase using established Class Rules of material testing, proof loading, and sample break loading. Assurance activities for component strengths are typically visual inspections, and other inspection techniques including chain measurements and should be considered for critical segments, such as top chain in the splash zone area.

*Fatigue life*—The performance of the fatigue life of a mooring line is also indicated by the safety factor. The minimum required fatigue life of the mooring system and the corresponding safety factor should be given in the performance parameter. While fatigue damage is not an item that can be readily measured, it can be calculated taking into account corrosion, wear, and loading history. A monitoring system with the capability of reporting tension time-series can facilitate in estimating fatigue accumulation. However, it is very difficult or nearly impossible to get early warning of fatigue failure due to many variables that can influence fatigue life, including the large variability caused by the scatterings associated with SN or TN data.

## 14.1.2   Assessing hazards and performing risk assessment

There are many hazards (or failure modes) associated with a mooring system. Specific hazards commonly seen are:

- Strength—overloading, low safety factor, snatch loads.
- Fatigue—tension, tension bending, torsion, out-of-plane bending.

- Corrosion/pitting—galvanic, splash zone, sulfate-reducing bacteria.
- Wear/erosion—interlink grip, chain stopper, hawse pipe, shackle pin, seabed.
- Clashing/contact—dropped object, off-take tanker, nearby vessel, work wire from a vessel.
- Manufacture defect/brittle fracture—toughness below spec, imperfection.
- Improper design—missing spacers in H-links, wire/polyester ropes in touchdown zone.

To determine whether a risk review or a risk assessment should be carried out for an existing mooring system, the following factors can be considered:

- Any severe weather events.
- Reported problem of a similar mooring system or component.
- Installation of new structures nearby or subsea.
- Modification of the design or adding novel component.

Each operator may have accumulated limited data from the fleet of their floating facilities. The industry as a whole can benefit from a comprehensive mooring failure database available in the public domain. The database can serve as an input for estimating failure probability and consequences. There are a few technical papers [5−7] that have compiled a collection of data for a specific range of years. Those data may serve as inputs or references for a risk assessment. A full quantitative risk assessment may not be easy to conduct, since both probability of failure and consequence of failure are difficult to estimate. In the current practice, a qualitative or semiquantitative risk assessment may be suitable for reviewing mooring system risks [2,3] by using a risk matrix.

When a high-risk item is identified in a mooring system, risk reduction measures can be taken, such as increasing inspection frequency, early detection with a monitoring system, replacing degraded components, or modifying the design preemptively. Monitoring plays an important role in the MIM process and is discussed in some detail in the next section.

## 14.2    Incident response

When an adverse event does happen, such as a potential line break, the operator should know how to respond in a prepared manner. Every permanently moored facility should have an incident response plan, commonly known as a Mooring Rapid Response Plan (MRRP) in place, which lays out a sequence of steps that should be followed in the event of a mooring line failure on a floating production facility. The MRRP can help operators to assess whether production can be safely continued, and to restore mooring integrity as efficiently as possible. The goal is to manage the risks associated with a mooring line failure, and thus MRRP is also referred to as Mooring Risk Management Plan [5]. These risks may be related to personnel safety, facility safety, oil and gas production, and/or environmental safety.

## 14.2.1 Define response actions

To effectively respond in the event of a mooring failure, the response plan should have a set of predetermined response procedures to follow. The response procedures may include up to three separate phases, each with its own associated time frame relevant to the incident [5]. The recommended procedures in each phase describe the major considerations for key personnel, and main activities to be achieved. The three phases [5] are described below:

### 14.2.1.1 Phase 1—emergency response

The emergency response should summarize the actions to be performed during the first 12 (or 24) hours of a mooring line failure.

- Identify and Verify the Mooring Line Failure—The first thing the crew should do is confirm that a mooring line has truly failed and that the failure was not a false alarm. During the mooring line failure confirmation, the failed component (e.g., mooring chain, polyester rope, shackle, etc.) should be identified. A mooring line failure can be confirmed in multiple ways, including a visual inspection, vessel position, or line tension measurements. Each vessel mooring system is different, therefore, it is important to clearly document the mooring line failure confirmation method to be used within the response plan.
- Notify Key Personnel of Mooring Line Failure—The crew should notify the rapid response team and all key personnel of the mooring failure in accordance with the agreed upon personnel list.
- Determine Whether to Shut Down Production—The rapid response team should follow the predetermined shut-in criteria and perform the following additional assessments to decide if it is safe to continue production operations with the damaged mooring line.

### 14.2.1.2 Phase 2—condition assessment

- Initiate a Root Cause Analysis (RCA)—An RCA should be initiated following the mooring line failure to determine why the mooring line failed. Recovery and onshore storage of the failed component may be useful for the RCA. The RCA is important because if the cause is related to the fabrication process, other components could fail by the same failure mode as the initial failure.
- Assess Additional Damage—Upon failure, the broken mooring line may make contact with other mooring lines, risers, umbilicals, or subsea

equipment during the fall to the seabed. It is recommended that a full diver or remotely operated vehicle (ROV) survey be conducted to search for any additional damage that the failed mooring line may have caused. This survey should inspect all intact mooring lines and risers. Note that this survey does not replace regular inspection activities related to a mooring integrity program.

- Assess Risk to Continued Production—Loss of production over an extended period is something to be avoided if possible. However, it is important to assess the risks of continued production with one mooring line damaged; therefore, a safety assessment is needed. The response plan should describe a method and acceptance criteria for continued production with one-line damaged.

### 14.2.1.3   Phase 3—mooring repair

- Start Line-Replacement Campaign—Within the response plan, there should be a high-level installation procedure for the repair of the failed mooring line. It is important during this phase to finalize the installation procedure with the help of the offshore installation contractor.
- Execute the Repair—The mooring line replacement activity should take the size, complexity, and criticality of the campaign into consideration. Note that, in many cases, mooring line handling onboard the floating production system is needed. On some floating systems, the installation winch may have been completely removed after the original installation. It is important to ensure the readiness of the onboard line handling equipment.

## 14.2.2   Include a sparing plan

The response plan should summarize the sparing strategy and have a clear record of available inventory. In the event of a mooring failure, spare mooring component(s) will be required. It is important to have a sparing philosophy in place to minimize the time required to restore the mooring to its designed state. A common sparing philosophy is to have one complete set of spare components for a single mooring line. However, it needs to be noted that current standard practice in the industry is to keep a full mooring line spare only for installation. The full spare is then sold or scrapped, and not kept for operations.

Driven piles and suction pile anchors are typically very robust, and therefore, do not require a spare in the operation phase. (Note that a spare during the original installation phase is often required.) However, the design drawings and specifications should be readily available in case new driven piles or suction piles need to be fabricated. Identification of existing off-vessel spare mooring equipment (chain, polyester ropes, wire ropes, connectors) is important to determine the readiness for a mooring repair, if required. It may be possible that other facilities in the area have existing spares which could be used for replacement of the failed mooring line. This could be for a permanent or temporary repair. Note that chains, polyester ropes, wire ropes, connectors, and anchors typically have long lead-times.

### 14.2.3   Predefine installation procedures and contracting plan

The response plan should have an installation procedure prepared in advance. In the event of a mooring line failure, it is critical to have a repair plan in place that not only has the procedure for repair, but also information on who in the region can perform the work. A step-by-step mooring line replacement procedure(s) should be developed. When a mooring leg is out of service due to a line break or an anchor failure, the production may need to be shut-in until the root cause of the mooring line failure is determined. To avoid a shut-in, a temporary mitigation plan that can remedy the one-line damaged condition may be developed as a part of the response plan to allow for uninterrupted production. It should be noted that there is a high likelihood that the adjacent mooring lines may fail in a similar manner when the first failure was caused by manufacturing issues or degraded components.

### 14.2.4   Include procedures for readiness check of equipment

The response plan should assess the readiness of on-vessel mooring equipment. Some of the older floating systems may not have the winching/jacking equipment permanently installed onboard. The on-vessel mooring equipment, such as pull-in winches and hydraulic power units, may have been removed or may not be functional. This will become a problem when a mooring repair needs to be done quickly. When on-vessel mooring equipment is available, it is important to ensure that it is acceptable and functioning to aid in the replacement of the failed line in the event of a mooring failure.

## 14.3   Life extension

Floating production facilities are installed at locations with a specific design life usually aligned with the design field life. A typical design life is 20 years, though there are facilities installed with design lives greater or less than 20 years. Most of them are designed for uninterrupted operation onsite without any dry docking. When a facility approaches the end of its design service life, the owner/operator may desire to have it remain on its location and continue its production operation. In these instances, the owner/operator typically initiates a life extension process with a classification society or the local regulatory agency. An evaluation is to be made and appropriate actions are to be taken to extend the life up to the new operating life. This process includes a reassessment of the floating system (including structure, stability, marine systems, and other machinery) and its mooring system. This reassessment normally includes both engineering and survey activities.

### 14.3.1 Life extension for a floating facility and its mooring system

The general procedure for continuing or extending the service life of an existing floating system can be summarized as follows [8]. First, survey the hull structure, mooring, and risers to establish the current conditions. Second, review the results of the structural, mooring, and riser analyses, utilizing the results of survey, original plans, and the most up-to-date metocean data to confirm that all the design criteria are met. Third, make any required repairs and modifications. Once these activities are completed to the satisfaction of a Class and/or regulatory agency, a life extension may be approved and granted.

Surveying the existing mooring system is necessary to determine a baseline condition upon which justification of continued service can be made. The typical scope for mooring baseline survey may include the following:

- Carry out general visual inspection (GVI) of off-vessel mooring components from top chain to the top of anchor piles, following applicable sections of API RP 2I [9].
  - Inspect for cracks, corrosion, and wear; conduct dimensional checks as accessible. A reasonable length of each mooring chain is to be cleaned to ensure the overall condition of each chain can be satisfactorily verified.
  - Inspect wire ropes and synthetic ropes for mechanical damage, twist and sheathing conditions, and anodes on wire sockets if installed.
- Carry out GVI of on-vessel mooring equipment including fairleads, windlasses, chain jacks, and chain stoppers. Additional inspections such as close-up visual inspection and/or NDE (Nondestructive Examination) may be needed for suspect areas.
- Carry out GVI of turret bearings, bogies (wheelset), and wear pads. If accessible, bearings and races are to be inspected.
- Verify any significant elongation of polyester ropes by checking if there is any large reduction in the measured pretensions. A calculation may be needed to estimate the new length of the ropes. It can be done by increasing the rope length incrementally in a mooring software until the calculated pretension matches the field measurements.

In many cases, a mooring system can receive a life extension approval after engineering (reassessment) and survey activities that prove its fitness-for-purpose. In some cases, these activities may conclude that certain mooring hardware components need to be replaced or repaired [10,11]. For instance, a mooring system originally designed for 10 years in a marginal/small field may have unsheathed wire ropes in its make-up. To get a life extension for the mooring system, the unsheathed wire ropes with a short design life may need to be changed out. Fig. 14.2 shows an unsheathed wire rope that was getting changed out by an anchor handling vessel in the field [10]. A brand-new sheathed wire rope with a yellow jacket, as shown in the same Figure, was used to replace the aged ropes.

**FIGURE 14.2**  Unsheathed wire rope getting replaced by brand-new sheathed wire rope during a life extension campaign [10].

### 14.3.2  Fitness assessment of mooring component

The remaining strength and fatigue life should be calculated and assessed to ensure the fitness for the extended life. This assessment is sensitive to winds, waves, and currents encountered and operating loads during the past service and future prediction, and therefore the long-term environmental data are to be properly represented. Mooring system models are to be developed and updated, incorporating wastage in the model while assessing mooring strength and remaining fatigue life.

The mooring strength reassessment is performed for the design environmental conditions. If there is updated metocean data, the design environmental conditions should be updated accordingly. Corrosion allowance should be updated based on the actual measurements from the past inspection. Remaining strength of mooring components at the end of the extended life should be estimated and checked against the required safety factors.

In calculating the accumulated fatigue damage or used-up fatigue life, the original safety factors may be reduced, providing that the technical justifications submitted by the owner/operator reflect a reduction of the uncertainty in the original design. The reduction of the uncertainty may be supported by the following evidences [8]:

- No past findings of fatigue cracks;
- No sharp contour on surface of chain links caused by corrosion or pits; and
- Having revised metocean data that are reliable and accurate.

An alternative methodology is to conduct fatigue testing on a chain retrieved from one of the mooring legs in the field [12]. It may be the most reliable method to prove the fitness-for-service for the life extension, but the cost of replacing and retrieving a segment of chain can be extremely high. Therefore, such an approach is seldomly used by operators.

Sheathed steel wire ropes and polyester ropes are less of a fatigue issue than chain due to fatigue lives that are much longer than required. In situ ROV inspection of the wire or polyester ropes may be considered by checking for any anomaly. Where turret mooring systems are installed, mechanical components such as turret bearings, swivel seals, and driving arms/mechanisms may need to be checked for fatigue.

When changes in design or load have been identified, the mooring system is to be reassessed. These changes may include mooring component modifications, metocean condition updates, and any other possible changes affecting the mooring responses.

Advanced analysis with FEA (finite element analysis) can be used to approve or reject assessment of the remaining strength of corroded chain links. It can be a cost-effective alternative to physical tests. Fig. 14.3 shows the process for creating the mesh surface of a corroded chain link. With a surface mesh of a high resolution and a stress plot [13] as shown in Fig. 14.4, FEA can predict the residual strength of a degraded chain link to a high degree of accuracy [14−17].

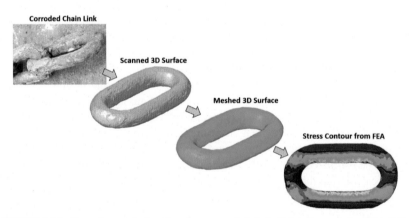

**FIGURE 14.3** Preparing meshed surface for a corroded chain link before running FEA [13]. *FEA*, Finite element analysis.

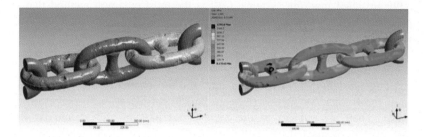

**FIGURE 14.4**   Model of corroded chain links for FEA (left); stress and deformation plot (right) [14].

While physical tests can be costly, a limited number of strength and fatigue test data on corroded chain links are available in the literature. Some of them were summarized in papers published based on findings from JIPs such as SCORCH [14,18] and FEARS [15], and others were the research results published by operators [12]. These published research results demonstrate that the technology for life extension assessment has been advancing, and will provide guidance to and raise the confidence level on any mooring life extensions in the future.

## 14.4   Ways to improve mooring integrity

Facing a high occurrence of potential issues and failures, mooring engineers should examine ways to improve mooring integrity. During the manufacturing phase, it is important to enforce QA/QC (Quality Assurance/Quality Control) on the mooring components. After the mooring system is installed, a combination of in-service inspection and monitoring can assess the condition of a mooring system and identify the trends of deterioration. Sharing lessons learned needs to be encouraged among owners and operators so that codes and standards can be updated and improved. These are the potential ways to improve mooring reliability [6].

### 14.4.1   Perform rigorous inspection and maintenance

Mooring components are designed to allow for some wear and corrosion, and inspection is to confirm that the wear and corrosion is within the allowable values over the design life. Baseline data, such as chain size when manufactured, are vital to assess how much wear and corrosion is occurring in the field. This enables estimates to be made on whether the chain is still fit for future service. During manufacturing phase, it is important to enforce QA/QC requirements and perform a rigorous inspection on the mooring components going through manufacturing.

For installed mooring systems, GVI is the most common method that is carried out by a slow ROV flight past the components being inspected. It is used to assess the overall condition of the mooring legs and to determine if any further inspection required. For shallow-water mooring systems, divers are often used.

Along the length of a mooring line, the top chains in the splash zone and at the fairlead/stopper have been the susceptible areas to corrosion or fatigue. They are subject to the highest degradation and should be the most closely inspected. Although time-consuming, cleaning of marine growth in key areas should be undertaken to ensure that defects are not missed during the GVI. Where there is a weight discontinuity in a mooring line, such as at a connector between chain and wire rope, the area may experience increased relative rotation/motion, which causes additional bending and wear. Wire rope terminations and connector areas both suffer from this. Also, experience has shown that the dynamic motion and wear seem to be particularly pronounced on leeward lines (i.e., the least loaded lines) [4]. In summary, critical areas to inspect can be categorized into the following four:

- Top chains in way of the splash zone, fairleads, and stoppers.
- Wire rope terminations.
- Seabed touchdown area.
- Connectors such as joining shackles and tri-plates.

Refer to Chapter 12, Inspection and monitoring, for a detailed review on inspection of mooring components.

Mooring standards and Class Rules are largely based on experience gained on temporary systems where smaller diameter chains could be inspected in dry conditions on deck during mobile offshore drilling unit (MODU) moves on a regular basis. For permanent mooring systems, more rigorous inspection practices should be implemented to account for the lack of accessibility to the mooring components. Note that permanent anchors are uninspectable by design and so is the buried section of anchor chain. Normally, they don't experience corrosion issues due to the fact that they are buried and thus blocked from oxygen.

Special attention should be given to the early years of service life. Using updated data based on recent papers [6,7], Fig. 14.5 has been created to show the trend along field life. It can be seen that there is a clear trend of "Infant Mortality." More than half of the incidents happened during the first 5 years of their design lives. More incidents occurred in the very first year, infant stage, than any other years. Following this observation, it may be recommended to set up an enhanced inspection program for the first few years with higher frequency or larger scope.

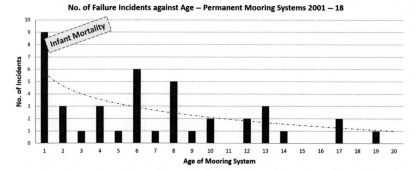

**FIGURE 14.5**   Number of incidents versus age showing infant mortality.

In terms of maintenance, mooring components subject to rough service conditions should be retrievable for inspection and maintenance on a periodic basis. Examples are chain link in a stopper or wire/chain on a fairlead. Chain links may be shifted in or out to move the fatigue and wear spots. Ultimately, they can be replaced if they are no longer fit. This maintenance practice can significantly improve the durability of the mooring legs. It should be taken into account during the design process.

### 14.4.2   Equip with monitoring system

Every floating system should have a monitoring capability to assure that it is safely moored. There have been many floating production vessels that have no means of knowing whether their mooring systems are intact. This is an undesirable situation, and such practice needs to be improved. Some of the newer mooring systems are equipped with monitoring devices capable of measuring mooring line tensions and displaying data continuously. Such systems can provide an instantaneous warning of mooring line failure, and record the tension history for later investigation. While most Class Rules do not require tension monitoring systems, operators should consider them in their design specifications. The cost of these systems is not substantial if included during construction, and they can reduce the risk of pollution and production shutdown.

Monitoring is an essential part of MIM. In some past incidents, mooring failures have gone undetected for months. In one case, the crew on an FPSO stopped paying attention to the monitoring system due to its poor user interface, and a line failure went unnoticed for several months until a diver spotted it while doing other work. Early detection of a single-line break can prevent multiline failure, and thus improve mooring integrity. It is important for all floating facilities to have a mooring line monitoring system such that a mooring line failure can be detected quickly. Refer to Chapter 12, Inspection and monitoring, for a detailed review on mooring monitoring.

### 14.4.3 Share lessons learned

As discussed in the previous chapter, novel failure mechanisms have occurred that may not be incorporated into industry design standards. As mooring designs progress into deeper water and harsher environments, and use component sizes and materials beyond the range of current experience, the occurrence of unanticipated failures due to newly discovered failure mechanisms will likely continue. Therefore, moorings that are designed according to current and accepted practice may not be as reliable as intended. In order to minimize the effects of pushing mooring designs beyond what the design codes are based on, it is important that operators and designers share lessons learned as early as possible. Design codes and guidance notes need to be developed to reflect some of the latest experiences, especially to capture new design criteria that are essential in coping with these novel problems.

Technical papers are also an important means of providing information and background on existing systems and mooring failures. There has been improved openness in the industry in discussing and addressing mooring related issues. It has resulted in JIPs and forums where experiences and opinions are shared [19,20]. A database of mooring issues and failures can be an effective tool in documenting the root causes of incidents and preventing them from reoccurring. It is a commendable practice where some safety alerts [2,3,21−24] are issued by operators or regulatory agencies. Those have created awareness and a positive impact in improving mooring integrity.

Some feedback is provided in Ref. [24] on how past experience can be captured in new designs. The same reference also provides some guidance on building redundancy or margin in new designs to provide allowance over the minimum requirements of Class or codes. From a practical perspective, it may be more cost-effective to build redundancy in a new design during the capital expenditure (CAPEX) phase compared to a mooring repair or replacement in the future, especially if the facility is still producing. However, this approach is often challenged at the CAPEX phase of the project where a major driver is to minimize cost.

### 14.4.4 Improve codes and standards

As the offshore industry starts to moor MODUs and permanent floating production units in harsh-weather areas or environmentally-sensitive areas, the risk of a mooring system failure becomes even higher. It is understandable that stakeholders may want to see higher design criteria to be adopted in mooring designs. The question is how to improve codes and standards so that the reliability of mooring systems can match that of other systems or other industries.

A common perception is that higher mooring system design criteria would result in stronger mooring lines, thus leading to greater survivability of the mooring system or lower probability of failure. The flaw of this perception is that the integrity of a mooring system is not solely controlled by its design strength. There are many factors that affect a mooring line's integrity, such as its fatigue design, corrosion protection, or the manufacturing quality of its components. Therefore, just increasing the design sea state for the mooring system survival condition or using a higher factor of safety would not necessarily result in more reliable mooring systems. Pushing the design criteria higher would result in larger component sizes that may be beyond existing manufacturing and QA/QC capabilities and may compromise component quality. Examining the mooring failures listed in Table 13.1 in the previous chapter and their root causes, it is clear that many of the mooring failures were not due to weather overload, but due to other issues, such as hardware's manufacturing quality or corrosion issues.

The above observations point out that the priority for improving codes and standards may be to strengthen specifications for mooring components such as chain. Additionally, it will be very practical to provide recommended practice and updated guidance on in-service inspection of mooring hardware such as those in API RP-2I [9]. The essential point is to improve the mooring system integrity by imposing sufficient requirements on mooring hardware and putting equal emphasis on all processes of design, manufacturing, handling, installation, inspection, maintenance, and monitoring. To raise the bar for the future reliability of mooring systems, design codes and standards need to be regularly updated and improved.

## 14.5   Questions

1. Vessel offset is a key performance parameter for a mooring system. To manage the integrity of a mooring system, name two other performance parameters to track and evaluate.
2. Explain briefly what "infant mortality" means in the context of mooring integrity.
3. As a crew member on a floating production vessel, you have noticed that one of the mooring lines has broken. What are the best actions to be performed by you and your OIM (Offshore Installation Manager) during the first 24 hours?
4. You are pursuing a life extension for a mooring system. You have calculated the remaining strength of the mooring chain based on dimension measurements made from a recent inspection. What other calculation should be done to ensure the fitness of the mooring chain for the extended life?
5. Briefly describe a couple of potential ways to improve mooring integrity.

# References

[1] C. Carra, T. Lee, K. Ma, A. Phadke, D. Laskowski, G. Kusinski, Towards API RP 2MIM – DeepStar guidelines for risk based mooring integrity management, in: Deepwater Offshore Technology, Oct. 2015.

[2] DeepStar, Mooring integrity management guidelines, in: Prepared by AMOG Consulting, DeepStar Phase XI CTR 11405, August 2013.

[3] API RP-2MIM, Recommended practice for mooring integrity management, in: Final Draft for Ballot, 2018.

[4] Oil & Gas UK, Mooring Integrity Guidance, November 2008.

[5] S. Bhattacharjee, D. Angevine, S. Majhi, D. Smith, Permanent Mooring Reliability & Mooring Risk Management Plan (MRMP): A Practical Strategy to Manage Operational Risk, Offshore Technology Conference, OTC-25841-MS, 2015.

[6] K. Ma, A. Duggal, P. Smedley, D. LHostis, H. Shu, A historical review on integrity issues of permanent mooring systems, in: OTC 24025, OTC Conference, May 2013.

[7] E. Fontaine, A. Kilner, C. Carra, D. Washington, K. Ma, A. Phadke, D. Laskowski, G. Kusinski, Industry survey of past mooring failures, pre-emptive replacements and reported degradations for mooring systems of floating production units, in: OTC 25273, OTC Conference, May 2014.

[8] ABS, Guidance notes on life extension methodology for floating production installations, in: American Bureau of Shipping, July 2015.

[9] API RP-2I, Recommended Practice for In-Service Inspection of Mooring Hardware for Floating Structures, third ed., April 2008.

[10] K. Ma, et al, Life extension of mooring system for benchamas explorer FSO, in: Proceedings of the 19th Offshore Symposium, Society of Naval Architects and Marine Engineers (SNAME), Houston, TX, February 2014.

[11] M. O'Driscoll, H. Yan, K. Ma, P. Stemmler, Replacement of corroded mooring chain on an FPSO, in: Ship Production Committee, SNAME Maritime Convention, 2016.

[12] S. Fredheim, S. Reinholdtsen, L. Haskoll, H.B. Lie, Corrosion fatigue testing of used, studless, offshore mooring chain, in: Proceedings of the 32nd International Conference on Ocean Offshore and Arctic Engineering, OMAE 2013-10609, Nantes, France, June 2013.

[13] S. Wang, X. Zhang, T. Kwan, K. Ma, et. al., Assessing fatigue life of corroded mooring chains through advanced analysis, in: Proc. Offshore Technology Conference, OTC 29449, May 2019.

[14] E. Fontaine, A. Potts, K. Ma, A. Arredondo, R. Melchers, SCORCH JIP: examination and testing of severely-corroded mooring chains from West Africa, in: Proc. Offshore Technology Conference, OTC 23012, May 2012.

[15] J. Rosen, G. Farrow, A. Potts, C. Galtry, W. Swedosh, D. Washington, A. Tovar, Chain FEARS JIP: finite element analysis of residual strength of degraded chains, in: Proc. Offshore Technology Conference, OTC 26264, May 2015.

[16] J. Crapps, H. He, D. Baker, Strength assessment of degraded mooring chains, in: Proc. Offshore Technology Conference, OTC 27549, May, 2017.

[17] P. Vargas, T.M. Hsu, W.K. Lee, Stress concentration factors for stud-less mooring chain links in fairleads, in: Proceedings of the 23rd International Conference on Ocean Offshore and arctic Engineering, OMAE2004-51376, Vancouver, June 2004.

[18] J. Rosen, A. Potts, E. Fontaine, K. Ma, R. Chaplin SCORCH JIP – feedback from field recovered wire ropes, in: Proc. Offshore Technology Conference, OTC 25282, May 2014.

[19]   M. Brown, FPS mooring integrity JIP, in: Noble Denton, A4163, Rev. 1, Dec. 21, 2005.

[20]   HSE, JIP FPS mooring integrity, in: UK HSE Research Report 444, Prepared by Noble Denton, 2006.

[21]   MMS (BSEE), Catastrophic failures in mooring systems possibly put floating structures at risk, in: Safety Alert No. 259, United States Department of the Interior, January 2008.

[22]   BSEE (MMS), Accident Investigation Report — 15 May 2009, United States Department of the Interior, Minerals Management Service, July 2010.

[23]   BSEE (MMS), Catastrophic failures in mooring systems possibly put floating structures at risk, in: Safety Alert No. 296, United States Department of the Interior, May 2011.

[24]   A. Duggal, W. Fontenot, Anchor leg system integrity — from design through service life, in: Proc. Offshore Technology Conference, OTC 21012, May 2010.

# Chapter 15

# Mooring for floating wind turbines

## Chapter Outline

15.1 **Concepts of floating offshore wind turbines** **300**
    15.1.1 History of concept development 300
    15.1.2 Spar-buoy type 301
    15.1.3 Semisubmersible type 301
    15.1.4 Tension leg platform type 302
    15.1.5 Comparison of concept types 303
15.2 **Mooring design** **304**
    15.2.1 Mooring type 304
    15.2.2 Mooring line material 305
    15.2.3 Anchor selection 306
15.3 **Mooring design criteria** **307**
    15.3.1 Design return period 307

15.3.2 Optional redundancy 307
15.3.3 Other requirements 308
15.4 **Mooring analysis** **308**
    15.4.1 Environmental forces and load cases 308
    15.4.2 Aerodynamic loads 310
    15.4.3 Time-domain mooring analysis 311
15.5 **Design considerations** **312**
    15.5.1 Fatigue 312
    15.5.2 Corrosion 313
    15.5.3 Installation 313
    15.5.4 Tensioning 313
    15.5.5 Overall project cost 313
15.6 **Questions** **314**
**References** **314**

Driven by the development of renewable energy, floating offshore wind turbines (FOWTs) have been developing rapidly over recent years. The mooring system design of FOWTs has played an essential role in their feasibility. The analysis and design of such mooring systems are a natural extension from the practice of the offshore oil and gas industry. In this chapter, the types of FOWTs and their associated mooring systems are reviewed. Their design criteria are summarized and their differences from the traditional mooring systems for oil and gas productions are highlighted.

Mooring System Engineering for Offshore Structures. DOI: https://doi.org/10.1016/B978-0-12-818551-3.00015-6

## 15.1 Concepts of floating offshore wind turbines

### 15.1.1 History of concept development

Early offshore wind farms were built in shallow waters of less than 40-m depth. Fixed monopiles (or jackets) were used as the supporting structure. As the wind farm development moved into areas of water depth greater than 50 m, fixed monopiles became costly and the FOWTs became the preferred solution.

The first scaled prototype, Blue H, was installed off the coast of Italy in 2008 in a water depth of about 113 m [1]. Since then, several prototypes and full-scale FOWTs have been installed and deployed for concept demonstrations, including the Hywind FOWTs installed offshore Norway in 2009 [2], and the WindFloat FOWT installed offshore Portugal in 2011 [3]. In 2017 Hywind Scotland was completed and became the first commercial floating wind farm, with five floating turbines with a total capacity of 30 MW [4].

As of 2019, there are over 30 floating wind concepts under development [5–8]. Many of them leveraged the experience from the oil and gas industry. Each concept is designed for certain water depth, seabed conditions, local infrastructure, and supply chain capabilities. Like the floating systems for the oil and gas industry, FOWTs can also be categorized into three main types [7]: spar-buoy, semisubmersible, and tension leg platform (TLP), as illustrated in Fig. 15.1. There are also hybrid types of floating wind turbines, such as a spar and TLP combination type. Besides the three types, a barge-type hull can also serve as the floating foundation. A barge-type floater demonstrator, Floatgen (Damping Pool), has been designed and installed by Ideol in France.

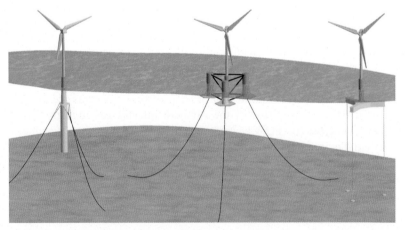

**FIGURE 15.1** Three types of floating offshore wind turbine floaters: spar, semisubmersible, and TLP. *TLP*, Tension leg platform.

## 15.1.2   Spar-buoy type

The spar type offshore wind turbine comprises the floating foundation, the tower, the rotor-nacelle assembly, and the mooring system. The floating foundation consists of a steel (or concrete) cylinder filled with a ballast of water and gravels to keep the center of gravity well below the center of buoyancy. It ensures the wind turbine floats in the sea and stays upright. The draft of the floating foundation is usually larger than (or at least equal to) the tower height above the mean sea level. The floating foundation can be towed in the horizontal position to sheltered waters. It is then upended and stabilized. The tower and the rotor-nacelle assembly are then mounted by a dynamic positioning crane vessel, before finally being towed in the vertical position to the deployment site for connection to the mooring system. Equinor Hywind is one of the pioneers in spar type FOWTs as illustrated in Fig. 15.2. Other spar-based concepts include, for example, Sway and Advanced Spar.

**FIGURE 15.2**   Spar-based wind turbine floaters. *Courtesy of Equinor.*

## 15.1.3   Semisubmersible type

Among the different types of FOWTs, the semisubmersible concept is the most versatile. The floating foundation comprises a few large columns connected by tubular members. The columns provide the stability for the floater, and therefore the semisubmersible type is also known as the column-stabilized type. A wind turbine may sit on one of the columns, or alternatively, may be positioned at the geometric center of the columns and supported by lateral bracings. A semisubmersible has a few advantages. One primary advantage is that it can be fabricated onshore in controlled settings and towed to the offshore site eliminating the need for an expensive construction vessel with a marine crane. When needed, the semisubmersible platform can be disconnected from its moorings and towed to a shipyard for maintenance. The relatively shallow draft allows for deployment at sites in very shallow waters. An example of a semisubmersible FOWT is WindFloat developed by Principle Power Inc. which consists of three columns with water-entrapment heave plates at the column bases, as shown in Fig. 15.3. Other semisubmersible-based concepts include Fukushima Shimpuu and SeaReed.

**FIGURE 15.3** Semisubmersible-based wind turbine floaters. *Courtesy of Principle Power.*

### 15.1.4 Tension leg platform type

The TLP type comprises a floating foundation (platform) to carry the wind turbine as shown in Fig. 15.4. Unlike the spar type which needs to be assembled offshore, this TLP wind turbine may be assembled and commissioned

**FIGURE 15.4** TLP-based wind turbine floaters. *TLP*, Tension leg platform. *Courtesy of the Glosten Associates.*

onshore, thereby avoiding the logistical difficulties of offshore assembly. The floating platform is held in position by vertical tendons (also called tethers) which are anchored either by suction piles, driven piles, or a template foundation. The pretensioned tethers provide the righting stability. A TLP wind turbine has been installed off the coast of Puglia, southern Italy by Blue H Technologies [1]. Apart from PelaStar developed by Glosten and Blue H TLP by Blue H Group, there are other TLP-based concepts, including Eco TLP and GICON-SOF.

### 15.1.5 Comparison of concept types

The main features of the above three types of FOWTs are briefly summarized in Table 15.1.

**TABLE 15.1** Concept comparison of floating offshore wind turbines.

| Types of floater | Semisubmersible | SPAR | TLP |
|---|---|---|---|
| Advantages | 1. Onshore assembly at quayside<br>2. Good for a range of water depth<br>3. Uses only tug boats to install<br>4. Easy to disconnect and tow for maintenance | 1. Excellent stability<br>2. Simple design and fabrication<br>3. Towed in vertical position | 1. Excellent heave motion<br>2. Onshore assembly at quayside<br>3. Compact hull size<br>4. Small foot print |
| Disadvantages | 1. Large ballast increases displacement<br>2. Relatively large motion | 1. Requires a crane vessel to assemble offshore<br>2. Deep draft requires deeper water | 1. May lose stability upon tether failure<br>2. Hard to disconnect<br>3. Requires suction or driven piles |

*TLP*, Tension leg platform.

## 15.2 Mooring design

A FOWT mooring system must restrain the vessel excursion and motion within certain allowable limits. In shallow waters, the allowances are usually governed by: (1) the bending restriction of the export electrical cable; and (2) the turbine acceleration limitation due to vessel motions, particularly pitch and roll. These require the mooring system to meet certain specifications, which can be the main focus in most of the FOWT mooring designs. Design improvements may be achieved by several means, such as adding clump weight to the touch down chains, using parallel ground chains with tri-plates, inserting lightweight synthetic fiber ropes to increase the geometry stiffness, etc. While the mooring design practice is briefly described in this section, more detailed discussions can be found in Chapter 4, Mooring design.

## 15.2.1 Mooring type

The design of the FOWT mooring systems is a natural extension from the practice of the offshore oil and gas industry. The mooring types used by existing FOWT prototypes and design concepts include the following:

- Spread mooring system with catenary (or taut) lines (Fig. 15.5).
- Tension leg system.
- Single-point mooring.

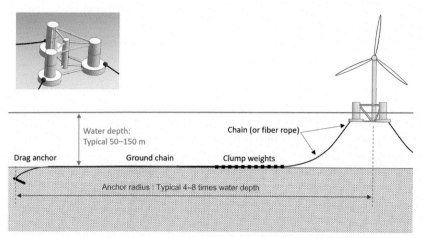

**FIGURE 15.5**  Typical spread mooring design with catenary lines.

For spar platforms, such as Hywind, the mooring system may comprise (3 × 2, three clusters with two lines per cluster) six legs for a design with redundancy or (3 × 1) three legs for a design without redundancy. The spar FOWT can be kept in position by a catenary or a taut mooring system. The make-up of a line can be chains, steel wire ropes, synthetic fiber ropes, or a combination of these.

For the three-column semisubmersible, such as WindFloat, an asymmetric mooring pattern may be applied. Two mooring lines can be connected to the column that carries the wind turbine, and one mooring line is connected to each of the remaining two columns [9].

### 15.2.2  Mooring line material

As the majority of FOWTs are expected to be deployed in shallow waters with a depth of less than 100 m, the catenary mooring system can adopt the "all chain" design that is commonly used in the oil and gas industry. The catenary shape and the weight of ground chain will provide the station-keeping function and keep the FOWT at its location. The catenary mooring leg has the ground chain resting on the seafloor to provide the restoring forces when getting lifted up by the vessel motion or excursion. Fig. 15.5 shows a typical mooring design with catenary lines in shallow water.

In waters deeper than 200 m, it is also possible to use a taut-leg (or semitaut) mooring design for floating wind turbines. In order to provide the compliance to floater dynamic responses, it may be a good option to incorporate synthetic fibers. The taut-leg system has the advantage of a smaller mooring footprint, but requires certain anchor types that can withstand the uplift force. It also provides a better station-keeping performance than a catenary mooring system. However, because FOWTs carry electrical cables rather than risers, the requirement on vessel (platform) offset is less critical than for oil and gas applications.

Synthetic fiber ropes can be a promising solution. They may be used in the line composition. Available materials are high modulus polyethylene (HMPE) (e.g., Dyneema) and polyester. Synthetic ropes have a long track record in the oil and gas industry, particularly in deep and ultradeep water with taut and semitaut configurations. They have the potential to deliver cost savings compared to conventional steel components due to their lower weight, better fatigue performance, and less dynamic tension due to their low stiffness. Details about these materials can be found in Chapter 9, Hardware—off-vessel components.

### 15.2.3  Anchor selection

The selection of anchors for FOWTs follows the proven practices used extensively in the oil and gas industry. There are a number of anchoring solutions available. The selection can be made based on the mooring configuration, seabed soil conditions, and holding capacity requirement. Applicable anchoring systems include:

- drag embedment anchor
- vertically loaded anchor (VLA)
- gravity installed anchor, such as torpedo anchor
- driven pile
- suction pile

Since the majority of FOWTs are expected to be deployed in shallow waters, a catenary mooring system is likely to be employed rather than a taut-leg system. Catenary mooring configurations commonly use drag-embedded anchors. Other anchor types may be chosen if the cost is competitive in the local market. However, the cheapest anchor choice is most likely the drag embedment anchor.

The taut-leg mooring systems typically use VLAs, driven piles, suction piles, or gravity installed anchors to cope with the vertical loads applied by the mooring lines. Piles can be used in difficult soil conditions. For TLP-based concepts, suction or driven piles will be the anchor choice because of the amount of vertical loads required. Piles are also the choice for anchor sharing. Since multiple FOWTs are usually installed at the same site, they can be strategically positioned so that their mooring legs can share anchors.

By using such a strategy, pile anchors will become more cost-effective. More details for anchor selection can be found in Chapter 8, Anchor selection.

## 15.3    Mooring design criteria

The design of offshore mooring systems has to comply with coastal country regulations and industry standards. Additionally, the operator often chooses to follow class society rules as well. There are a few existing standards that specifically address the design of FOWT hull structures and mooring systems. For example, class societies such as American Bureau of Shipping (ABS), DNV GL, and Bureau Veritas have developed specific rules governing the design of FOWT moorings [10−12]. There is also an international standardization effort for wind turbine installations led by the International Electrotechnical Commission (IEC). IEC is a worldwide organization that prepares and publishes international standards for electrical related technologies. It has released the standards for bottom-fixed offshore wind installations [13,14], and is publishing the standards for floating installations in 2019.

### 15.3.1    Design return period

The design of the moorings for FOWTs can follow the same standard as for the moorings of offshore oil and gas facilities with two main exceptions [15]. First, an oil and gas platform is usually a manned facility, while a FOWT is unmanned. They require different approaches to platform access, personnel protection, and safety. Second, the detrimental effect (consequence) to the environment due to a FOWT mooring failure is lower than a similar incident for floaters in the oil and gas industry. In other words, a FOWT does not handle hydrocarbons or hazardous chemicals, and thus presents significantly less risk to the environment in the case of failure.

Based on the two exceptions explained above, the return period specified for FOWTs is lower than that for the oil and gas facilities. For permanent floating production facilities, the 100-year environmental conditions are applied for the mooring design. For FOWTs, the 50-year environmental conditions are usually used.

### 15.3.2    Optional redundancy

The mooring systems for the oil and gas industry are required to have redundancy. In the event of one-line failure, the damaged mooring system must still be able to withstand the extreme design environment. For the moorings of FOWTs, the redundancy requirement can be waived by maintaining an increased safety factor at a "penalty factor" of 1.2. In other words, the minimum strength safety factor is typically increased by 20% (from 1.67 to 2.0) for the mooring intact condition, if the mooring system is not designed for a one-line damaged condition.

In short, the mooring system redundancy for FOWTs is optional. As a result, the mooring system design may be optimized between the two design scenarios, that is, with and without redundancy. The final choice can be made based on the lowest overall costs.

### 15.3.3  Other requirements

The fatigue safety factors in the class rules may also differ from those for offshore oil and gas applications. In general, slightly lower fatigue safety factors are allowed for moorings of FOWTs for the same reasons discussed above.

The offset of a FOWT is usually driven by the electrical cable design requirement. Depending on the water depth and cable configuration, a FOWT carrying electrical power cables can tolerate a larger vessel offset than an oil and gas facility carrying risers.

FOWTs are often close to coastal cities with large population. Due to the potential impacts to the fishing industry and marine wildlife, they are subjected to environmental evaluation. This needs to be taken into consideration for shallow-water moorings that tend to have large footprints. Also, some potential fields for FOWTs are situated in seismically active areas that could impact the anchor holding capacity. This needs to be taken into consideration during anchor design.

## 15.4  Mooring analysis

### 15.4.1  Environmental forces and load cases

The environmental load calculation in the mooring analysis for FOWTs is similar to that for the oil and gas platforms. In both cases, the moorings are subjected to the direct wind, waves, and current loads acting on the floaters as well as the additional loads caused by floater's motions (see Fig. 15.6 for illustration).

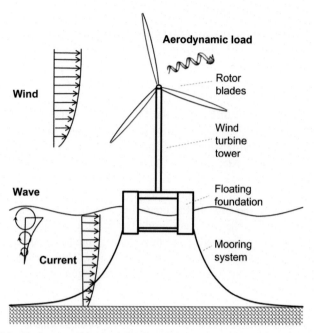

**FIGURE 15.6** Environmental loads on floating offshore wind turbines.

Based on IEC 61400-3 [14] as well as class rules [10], the mooring design conditions are to be represented by design load cases and survival load cases.

Design load cases are defined to verify the design adequacy of the FOWTs that are subjected to the combination of turbine operational conditions, site-specific environmental conditions, electrical network conditions, and other applicable design conditions. All relevant design load cases with a probability of occurrence are to be considered in the design. Combinations of these loads as well as the turbine operating conditions that produce the most unfavorable local and global effects on the mooring systems should be addressed.

Survival load cases under the 50-year environment are defined to verify the survivability of the mooring system, when the FOWTs are subjected to extreme environmental conditions. ABS [10] requires survival load cases to be checked under two conditions: (1) parked rotor-nacelle assembly with intact blades on the intact hull and mooring system; and (2) parked rotor-nacelle assembly with damaged blade(s) on the intact hull and mooring system.

### 15.4.2 Aerodynamic loads

One major difference between FOWTs and the floating systems used in the oil and gas industry is that the former subjects a complicated wind load that is affected by the aerodynamic effect of wind turbines. For a mooring analysis, the wind load can be determined using analytical methods or wind tunnel tests that use a reduced-scale model of the FOWTs.

One feature of wind turbines is the blade pitch, which refers to turning the angle of attack of the blades into or out of the wind. Due to the motions of the FOWTs, the inflow wind speed and angle relative to the turbine blades are coupled with the floater motion. The blade pitch is controlled by the control system [16]. The control system typically includes the generator-torque controller which maximizes the power capture for wind speeds less than the operating limit, and the blade-pitch controller which regulates the generator for wind speeds above the operating limit. The control system directly affects aerodynamic loads on the blades and thus impacts mooring line tensions and should be considered in the mooring fatigue design. The aerodynamic loads are also coupled with the wind turbine blade deflection, which is illustrated in Fig. 15.7.

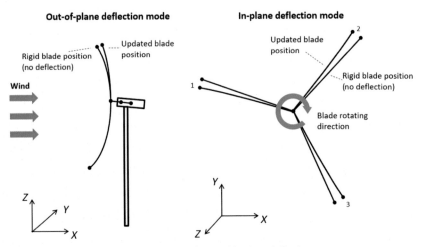

**FIGURE 15.7** Wind turbine blade out-of-plane and in-plane deflection.

Fig. 15.8 shows an example of an ideal turbine power output as a function of wind speed. In this example, the turbine is designed to reach full rated power at a wind velocity range of 12−30 m/s. Cut-in speed is the wind speed at which the wind turbine begins to produce power. The cut-out speed is the wind speed at which the turbine must be shut down to protect the rotor and drivetrain machinery from damage at high winds. This information is necessary to estimate the force and moment loading on the mast.

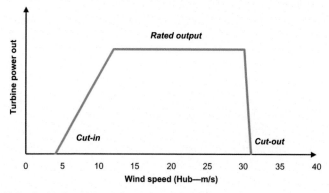

**FIGURE 15.8**   An ideal turbine rated power versus wind speed.

Based on IEC 61400-3 [14], as well as class rules [10], the possible turbine conditions should be considered, for example, start-up, power production, shut-down, parked (standing still or idling), and others. Aerodynamic loads induced by airflow passing through the rotor can be computed by the mean wind speed and air turbulence across the rotor plane, rotor rotational speed, air density, and aerodynamic shapes of wind turbine components. High-quality simulation tools are required when designing floating wind turbines because of the complex aerodynamic loads and the inherent dynamic response. There are a number of computer programs available to serve this purpose. One such tool, the FAST coupled aero-hydro-servo-elastic dynamic simulator, was developed and maintained by the US Department of Energy for use with offshore floating wind turbines [17]. FAST code has been incorporated into a number of computer software such as Orcaflex and SESAM for the design and analysis of offshore wind turbines [18–20].

For FOWTs installed in a wind farm, the potential shadow effect and wake effect on the wind load are to be considered for both the strength and fatigue analyses. For large wind farms, an increase in the turbulence intensity is to be taken into account in the wind load computation. As a general rule, the mutual influence of wind turbines through the wake interaction behind the rotor should be considered up to a distance of 10 times the rotor diameter.

### 15.4.3   Time-domain mooring analysis

Because interactions could occur among the rotor-nacelle assembly, the floating foundation, and the mooring system, the time-domain fully coupled dynamic analyses should be carried out to evaluate the dynamic responses of the FOWT [10]. The frequency-domain analyses cannot capture the nonlinear dynamic interactions among the components of the FOWT. Therefore most of the currently available simulation software for FOWT is based on the time-domain analysis. The frequency-domain analysis of floater motion is normally performed to calculate the hydrodynamic coefficients which are used as input for the time-domain analyses [10].

It should also be noted that floating wind turbines have smaller displacements compared to floating drilling or production structures, with the exceptions of catenary anchor leg mooring buoys. The former may have a total displacement of less than 15,000 t, while the latter typically has a displacement of around 50,000 t. In general, the FOWTs may exhibit more coupling between the moorings and the floating structure. Although to be assessed on a case by case basis, it is believed that a fully coupled analysis will make a better prediction of the behavior of the system. Prototype tests and model tests are recommended to supplement the mooring analysis. The state-of-the-art reviews of the design methodology and simulation software for FOWTs can be found in the references [5,8].

## 15.5    Design considerations

The standards for FOWT mooring designs are primarily based on the existing standards from the oil and gas industry. Although specific standards have been developed that address technical issues, such as design environment, mooring redundancy, and aerodynamics, there may be potential gaps to be closed in areas like integrity management and risk/reliability level [15,21]. The mooring systems designed for FOWTs are considered a recent development that may have opportunities for improvements. The following topics are a few technical areas that need more guidance and deserve further studies.

### 15.5.1    Fatigue

Fatigue is a major integrity threat for mooring systems [22]. Statistics from the oil and gas industry, as discussed in Chapter 13, Mooring reliability, show that fatigue is one of the main causes for mooring line breaks. The experience may suggest that mooring line failures can occur in FOWT moorings. In fact, FOWT will have livelier (i.e., more severe) motions than those floating production units, because the former is smaller in size and thus moves more actively in waves. The coupling of the aerodynamic load from the wind turbines introduces additional platform movements. All these platform motions will create cyclic tensions in the mooring lines and reduce the fatigue lives accordingly. In addition to the wave-frequency fatigue loads, the potential snap loads in extreme conditions can also adversely affect the fatigue lives. One potential mitigation is to use synthetic fiber ropes such as HMPE (e.g., Dyneema) or polyester to replace the top chain that is prone to fatigue.

### 15.5.2 Corrosion

Corrosion is another long-term integrity issue for mooring chain [23]. Generally, the corrosion problem is most severe in the splash zone and upper water column. One strategy to mitigate the corrosion problem is to place the fairleads deep below the water line, so the mooring chain is away from the splash zone. Another strategy is to use synthetic fiber ropes such as HMPE (e.g., Dyneema) or polyester to replace the top chain.

### 15.5.3 Installation

Mooring installation is a significant cost contributor for FOWTs. The fairlead connector (e.g., stopper or uni-joint) design and the selected anchor type have a significant impact on the installation method. Simple and cost-effective installation methods are needed to reduce the cost. The installation method for shallow-water mooring applications is fairly established. If drag embedment anchors are selected, the installation may conveniently reference any geotechnical data from nearby sites. The mooring lines are typically prelaid on the seabed and the top segment of the lines are tentatively suspended and held by marker buoys. When the floater is towed to site, it is connected to the prelaid mooring lines by an anchor handling vessel (AHV). Refer to Chapter 11, Installation, for typical installation procedures for permanent moorings.

### 15.5.4 Tensioning

Tensioning of mooring lines is one of the main tasks during installation. The conventional method is to place a temporary chain jack (or a winch) on the platform to tension up the mooring lines. Two alternative methods may have the potential to reduce the cost. One method eliminates onboard tensioning facilities and simply uses multiple tug boats to push the floater and allows the mooring line to be connected on an AHV with an H-link. Another method is to use an in-line tensioner. See Chapter 10, On-vessel equipment, for a detailed discussion on the use of in-line tensioners.

### 15.5.5 Overall project cost

The mooring designs adopted in early demonstration projects of FOWTs are based on traditional oil and gas practices. They may be conservative solutions. It is believed that there are opportunities for cost reductions through improved designs that minimize the capital expenditures and/or operating expenses.

## 15.6 Questions

1. Name at least two advantages for each of the FOWT floater types: spar-buoy, semisubmersible, and TLP.
2. Name one disadvantage for each of the FOWT floater types: spar-buoy, semisubmersible, and TLP.
3. Your engineering firm has been chosen to design a fleet of FOWTs which will be installed in 60-m water in the Taiwan Strait. As the chief naval architect, what floater type will you choose? Why?
4. How many mooring lines would you use for the particular floater type that you chose in the previous question? Why? For the strength of mooring lines, what is (are) the design safety factor(s)?
5. Why do most standards require a 50-year storm to be applied to FOWT moorings rather than the 100-year storm for permanent (floating production) moorings?

## References

[1] N. Bastick, 2009, Blue H—the world's first floating wind turbine, in: The First Dutch Offshore Wind Energy Conference, February 12 and 13, 2009, Den Helder, The Netherlands, Essential Innovations.

[2] S. Bratland, 2009, Hywind—the world first full-scale floating wind turbine, in: Seminar and B2B Meetings "Powering the Future—Marine Energy Opportunities", November 5, 2009, Lisbon, Portugal.

[3] A. Aubault, C. Cermelli, A. Lahijanian, A. Lum, A. Peiffer, D. Roddier, WindFloat contraption: from conception to reproduction, ASME 2012 31st International Conference on Ocean, Offshore and Arctic Engineering, American Society of Mechanical Engineers, 2012, pp. 847–853.

[4] Equinor, Hywind Scotland Brochure.

[5] ABS Report, Floating Wind Turbines, Technology Assessment and Research Program (TA&R Project No. 669), Bureau of Safety and Environmental Enforcement (BSEE), U.S. Department of the Interior, Washington, DC, 2012.

[6] ABS Report, Design Standards for Offshore Wind Farms, Technology Assessment and Research Program (TA&R Project No. 670), Bureau of Ocean Energy Management, Regulation, and Enforcement (BOEMRE), U.S. Dept of the Interior, 2011.

[7] Carbon Trust, Floating Offshore Wind: Market and Technology Review, Prepared for the Scottish Government, 2015.

[8] ABS Report, Design guideline for stationkeeping systems of floating offshore wind turbines, in: Technology Assessment and Research Program (TA&R Project No. 705), Bureau of Safety and Environmental Enforcement (BSEE), U.S. Dept of the Interior, 2013.

[9] D. Roddier, C. Cermelli, A. Weinstein, WindFloat: a floating foundation for offshore wind turbines—Part I: Design basis and qualification process, ASME 2009 28th International Conference on Ocean, Offshore and Arctic Engineering, American Society of Mechanical Engineers, 2009, pp. 845–853.

[10] ABS, Guide for Building and Classing Floating Offshore Wind Turbine Installations. American Bureau of Shipping, 2015, (Updated March 2018).

[11]  DNV, Design of Floating Wind Turbine Structures, 2013.

[12]  Burear Veritas, Classification and Certification of Floating Offshore Wind Turbines, 2015.

[13]  IEC 61400-1, Wind Turbines—Part 1: Design Requirements, 3.1 ed., 2014.

[14]  IEC 61400-3, Wind Turbines—Part 3: Design Requirements for Offshore Wind Turbines, 1.0 ed., 2009.

[15]  A. Aubault, D. Roddier, K. Banister, Regulatory Framework for Design, Construction and Operation of Floating Wind Turbine Platforms, OTC-27215-MS. Offshore Technology Conference, 2016.

[16]  J.G. Njiri, D. Soffker, State-of-the-art wind turbine control: trends and challenges, Renewable Sustainable Energy Rev. 60 (2016) 377−393.

[17]  J.R. Browning, et al., Calibration and validation of a spar-type floating offshore wind turbine model using the FAST dynamic simulation tool, J. Phys.: Conf. Ser. 555 (2014) 012015.

[18]  SINTEF Ocean, Simo Theory Manual, Version 4.12.2, 2018.

[19]  MARINTEK, Riflex User Manual, Version 4.12-02, 2018.

[20]  Orcina, OrcaFlex Manual, Version 9.7a., 2013. <https://www.orcina.com/>.

[21]  Carbon Trust, Floating Wind Joint Industry Project, Phase I Summary Report—Key Findings from Electrical Systems, Mooring Systems, and Infrastructure & Logistics Studies, 2018.

[22]  K. Ma, A. Duggal, P. Smedley, D. LHostis, H. Shu, A historical review on integrity issues of permanent mooring systems, in: OTC 24025, OTC Conference, May 2013.

[23]  K. Ma, Ø. Gabrielsen, Z. Li, D. Baker, A. Yao, P. Vargas, et al., Fatigue tests on corroded mooring chains retrieved from various fields, in: OMAE2019-95618, June 9−14, 2019.

# Index

*Note*: Page numbers followed by "*f*" and "*t*" refer to figures and tables, respectively.

## A

Aerodynamic loads, 310−311
Airy theory, 49
"All-chain" design, 175
American Bureau of Shipping (ABS), 307
American Petroleum Institute (API), 6−7
  Recommended Practice 2P (API RP 2P),
    6−7
American Petroleum Institute Recommended
    Practice (API RP) 2I, 234, 295
Anchor handling tug (AHT). *See* Anchor
    handling vessel (AHV)
Anchor handling vessel (AHV), 164, 217,
    219−221, 227, 229−230, 313
Anchor selection, 155
  anchor design considerations, 157
  available anchor types, 155−157
  drag embedment anchors (DEAs), 163−166
    advantages and limitations of, 164−165
    holding capacity of, 165
    installation and recovery, 165−166
  driven piles, 161−163
    holding capacity of, 162
    installation, 163
  gravity installed anchors, 170−173
    OMNI-Max anchor, 171−173
    torpedo anchor, 170
  soil characterization, 157−158
  suction embedded plate anchors (SEPLAs),
    168−169
    advantages and limitations of, 168−169
    installation, 169
  suction piles, 158−161
    holding capacity of, 159−160
    installation, 160−161
  vertically loaded anchors (VLAs), 166−168
    holding capacity of, 167−168
    installation, 168
    for permanent and temporary moorings,
      166−167

API RP 2FP1, 6−7
API RP 2SK, 6−7, 103−104, 167−168
AQWA, 52
Aramid (aromatic polyamide), 187−188, 190
Ariane, 111

## B

Barge-shaped hulls, 9
Beam-column method, 162
Blowout preventer (BOP), 3−4
Blue H, 300
Bottom-founded structures, 5
Boundary integral equation method
    (BIEM), 52
Bourbon Dolphin, 230−231
BTM (buoy turret mooring) system, 31, 31*f*
Buoy, 195−196
Bureau Veritas (BV), 104, 111, 307

## C

Capital expenditure (CAPEX), 206, 211, 294
Catenary Anchor Leg Mooring (CALM)
    buoy system, 15−16, 34−35, 34*f*, 177
Catenary equation, 87−89
Catenary mooring systems, 20, 20*f*, 306
Catenary versus taut leg moorings, 20
Cell spar, 12−13
Chain, inspection of, 241−243
Chain grades
  R3 grade chain, 94*t*, 177
  R3S grade chain, 94*t*, 177
  R4 grade chain, 94*t*, 177
  R4S grade chain, 94*t*, 177
  R5 grade chain, 94*t*, 177
Chain jack, 201, 203−204
  movable, 204−205, 205*f*
Chain locker, 202−203
Chain stoppers, 24, 201, 201*f*
Chain windlass, 204−206

Class Rules, 234, 283
C-link, 194, 194*f*
Close-end socket, on wire rope, 181*f*
Close-up visual inspection, 239−240
Clump weight, 196−197
Clustered moorings, 70−71
Clustered-spread mooring, 22*f*, 23
Cnoidal wave theory, 49
Column-stabilized type, 301−302
Compliant tower, 10
Comprehensive engineering analysis,
    155−156
Computational fluid dynamics (CFD), 52
Cone Penetration Tests, 158, 216
Connecter and anchor, 245
Constructional extension, 183
Corrosion, 241, 313
    protection, 182−183
Coupled analysis, 107−109
Coupled numerical model, 151
Current-induced forces and vortex-induced
    motion, 56−57

**D**

Decay tests, 140
DeepC/SESAM, 110−111
Deep-draft semisubmersible, 11, 12*f*
Delmar Systems, 171−172
Design basis, 64−65
    gathering input data, 64−65
Design considerations, 74−77
    avoiding clash or interference, 76−77
    limiting vessel offset, 75
    minimizing line tension, 75−76
    reducing fatigue damage accumulation, 76
Design criteria, 77−80
    design codes, 77
    fatigue design criteria, 79−80
    operability requirement, 80
    strength design criteria, 78−79
    vessel offset requirement, 77−78
Design load cases, 309
Design process, 65−74
    designing the mooring line composition,
        71−73
    designing the mooring pattern, 68−71
    determination of the profile, 67−68
    optimization, 73−74
    selecting the mooring system, 67
Det Norske Veritas (DNV), 177
Det Norske Veritas Germanischer Lloyd, 204

Devices, monitoring, 249−251
    global positioning system−based system,
        251
    inclinometer, 250−251
    load cell, 250
Differential Global Positioning System
    (DGPS) systems, 248−249, 251
Diffraction/radiation panel model, 54*f*
Direct tension measurement, 248
Disconnectable turret mooring system, 28−32,
    28*f*
Det Norske Veritas (DNV) criteria, 80
    DNV GL, 307
    DNV-RP-E302, 167−168
Drag coefficient, 95, 129
Drag embedment anchors (DEA), 156−157,
    163−166
    advantages and limitations of, 164−165
    holding capacity of, 165
    installation and recovery, 165−166
    typical configurations, 163*f*
Drilling riser, 8−9
Drillships, 3−4, 9
Driven piles, 161−163, 161*f*
    holding capacity of, 162
    installation, 163
    onshore methods for, 163
Drop anchors. *See* Gravity installed anchors
Drum winch, 206
D-shackles, 191, 192*f*
Dynamic positioning (DP) system, 35−38
    floating production storage and offloading
        with, 36*f*
Dynamic stiffness, 97−98

**E**

Eight-strand wire rope, 180*f*
Elastic stiffness, 90
Electric drive, 202
Emergency response, 285
Engineering analysis and code check,
    80−81
    load cases for, 82*t*
    mooring analysis load cases, 81
Environmental forces and load cases,
    308−309
Environmental loads and vessel motions, 41
    current load and vortex-induced motion,
        54−57
        current-induced forces and vortex-
            induced motion, 56−57

directional combination of wind, waves, and current, 60
ice load, 57–59
  description of ices, 58
  ice-induced forces and ice management, 58–59
loads on floating structures, 41–44
  loads in different frequency ranges, 43–44
  mooring system to resist environmental loads, 41–42
  site-specific environmental data, 43
sensitivity study on wave period, 60
wave–current interaction, 61
wave load and vessel motions, 48–54
  description of waves and swells, 48–51
  wave-induced forces and motions, 51–54
wind load, 44–47
  description of winds, 44–46
  wind-induced forces, 46–47
Equally-spread mooring, 23
Equally-spread versus clustered-spread moorings, 23
Equivalent (truncated) mooring lines, 148*f*
Equivalent mooring system, 148
"Evenly spread" moorings, 70–71
External disconnectable turret, 29
External turret mooring system, 24, 26–28, 27*f*

**F**
Failure mechanisms, 273–276
  chain damaged from handling, 276
  chain with severe corrosion, 274
  deficient chain from manufacturing, 274
  fatigued chain due to out-of-plane bending, 275
  knotted chain due to twist, 275–276
Failure probability ($P_f$) for permanent mooring systems, 262–263
Failure spots for permanent moorings, 264–265
Fatigue analysis, 115
  fatigue resistance of mooring components, 117–122
  comparison between T–N and S–N curves, 121–122
  S–N curves for chain and wire ropes, 119–120
  T–N curve for polyester ropes, 120–121
  T–N curves for chain, connectors and wire ropes, 118–119

in frequency domain, 122–125
  combined spectrum approach, 124–125
  dual narrow band approach, 125
  simple summation approach, 123–124
Miner's rule, 117
out-of-plane bending (OPB) fatigue for chain, 132–135
  mechanism of, 132–134
  OPB fatigue assessment, 134–135
procedure, 126–127
vortex-induced motion (VIM), 128–131
  mechanism of, 128–129
  VIM fatigue assessment, 129–131
Fatigue damage accumulation, 76
FEARS, 291
Fiber core, 181
Fiber rope, inspection of, 244–245
Fifth-generation drilling semisubmersible, 4*f*
Finite element (FE) method, 91
Finite element analysis (FEA), 135, 159, 290
Fixed platforms, 10
Floater motion analysis, 107
Floaters, modeling of, 93
Floating drilling, 7–10
  drilling semi, 8–9
  drillship, 9
  tender-assisted drilling (TAD), 9–10
Floating drilling vessels, 2, 8
Floating Liquefied Natural Gas, 14
Floating offshore wind turbines (FOWTs), 299
  comparison of concept types, 303
  concept development, 300
  design considerations for, 312–313
    corrosion, 313
    fatigue, 312
    installation, 313
    overall project cost, 313
    tensioning, 313
  mooring analysis for, 308–312
    aerodynamic loads, 310–311
    environmental forces and load cases, 308–309
    time-domain mooring analysis, 311–312
  mooring design criteria, 307–308
    design return period, 307
    optional redundancy design, 307–308
  mooring design for, 304–307
    anchor selection, 306–307
    mooring line material, 305–306
    mooring type, 304–305
  semisubmersible type, 301–302
  spar-buoy type, 301

Floating offshore wind turbines (FOWTs)
(*Continued*)
tension leg platform type, 302−303
Floating platforms, 1
Floating production, 5, 10−16, 70−71
Catenary Anchor Leg Mooring (CALM)
buoy, 15−16
facilities, 287−288
FPSO and FSO, 13−14
semisubmersible, 11
spar, 12−13
tension-leg platform, 11
Floating production unit (FPU), 2, 217, 255−256
Floating structure motion, 91−92
Floating structures, 41−44
loads in different frequency ranges, 43−44
mooring system to resist environmental
loads, 41−42
site-specific environmental data, 43
Floating wind turbine, 17−18
"Floating" drilling vessel, 2
Fluctuating wind, 45
Fluke, 163−164, 166−167
FPSO (floating production storage and
offloading), 1, 5, 13−14, 23−25
Fracture mechanics approach, 116
Frequency domain, 47, 102
fatigue analysis, 122−125
combined spectrum approach, 124−125
dual narrow band approach, 125
simple summation approach, 123−124
limitation of, 104−105
strength analysis, 102−105
frequency-domain analysis procedures,
103−104
response transfer functions, 102−103
Froude number, 147
Froude scale factors, 143*t*
FSO (floating storage and offloading), 13−14,
23, 33
Full-scale field tests, 51−52

**G**

Gantry structure, 24
General visual inspection (GVI), 237−238,
288, 292
Geometric stiffness, 90
Girassol FPSO, 206
Global positioning system, 251
Grain size distribution, 158
Gravity installed anchors, 170−173
OMNI-Max anchor, 171−173

torpedo anchor, 170
Grouped mooring systems, 70−71
Gulf of Mexico (GOM), 2−3, 6

**H**

Hardware−off-vessel components, 175
buoy, 195−196
chain, 176−180
chain grades, 177−178
manufacturing process, 178−180
studlink versus studless, 177
clump weight, 196−197
connectors, 191−195
connectors for temporary moorings,
193−195
for permanent moorings, 191−193
mooring line compositions, 175−176
polyester rope, 183−187
first use of polyester mooring in
deepwater, 185
polyester stretch, 187
rope constructions, 185−187
synthetic ropes, 187−191
aramid rope, 190
considerations for moorings in ultradeep
waters, 190−191
high modulus polyethylene rope, 188−189
nylon rope, 188
wire rope, 180−183
corrosion protection, 182−183
six-strand versus spiral strand, 181−182
termination with sockets, 183
Hazards, assessing, 283−284
High modulus polyethylene (HMPE) rope,
187−191, 306
H-links, 191−192, 192*f*
HPU. *See* Hydraulic power unit (HPU)
Hybrid model testing, 150
Hybrid test method, 150−151, 151*f*
Hydraulic drive, 201−202
Hydraulic power unit (HPU), 201−202
Hywind, 18, 300

**I**

Iceberg, 58
Ice load, 57−59
description of ices, 58
ice-induced forces and ice management,
58−59
Ice tank test, 142, 142*f*
Inclinometer, 248, 250−251

Independent wire rope core (IWRC), 181
Indirect tension measurement, 248
Industry practice, 107, 109
In-line tensioner, 209–212, 210f
Inspection, 233–234
    methods, 236–241
        advanced three-dimensional imaging,
            240–241
        close-up visual inspection, 239–240
        difference between mobile offshore
            drilling unit and permanent moorings,
            236–237
        general visual inspection, 237–238
        nondestructive examination (NDE)
            techniques, 240
    of mooring components, 241–245
        chain, 241–243
        connecter and anchor, 245
        fiber rope, 244–245
        wire rope, 243–244
    regulatory requirements, 234
    schedule, 234–236
        as-built survey for permanent mooring,
            235
        periodic surveys for mobile offshore
            drilling unit mooring, 236
        periodic surveys for permanent mooring,
            235–236
Installation, 215, 313
    deployment and retrieval of temporary
        mooring, 225–229
    preset mooring system for mobile
        offshore drilling unit, 228–229
    rig mooring system for mobile offshore
        drilling unit, 226–227
    installation vessel, 229–231
        anchor handling vessel (AHV),
            229–230
        capsizing of Bourbon Dolphin, 230–231
    of permanent mooring, 217–225
        Phase III—hook-up of mooring lines to
            floating production unit, 222–225
        Phase II—prelay of mooring lines on
            seabed, 219–222
        Phase I—installation of pile anchors,
            217–219
    site investigation, 215–216
        geophysical survey, 216
        geotechnical survey, 216
Integrity management of mooring systems.
        *See* Mooring integrity management
        (MIM)

Interlink stiffness, 133
Internal turret mooring system, 24–26
ISO 19901-7, 7

**J**
Jack-up barges, 7–8
Joint Industry Project (JIP), 188, 266, 291, 294
Joint North Sea Wave Observation Project
        (JONSWAP) spectrum, 50–51, 52f

**K**
Kenter link, 193–194, 194f

**L**
Level (and pack) ice, 58
Life extension, 287–291
    fitness assessment of mooring component,
        289–291
    for floating facility and its mooring system,
        288–289
Linear wave theory, 49–50
Linear winch, 209
Liquefied natural gas (LNG), 14
Load cell, 248, 250
Loads in different frequency ranges, 43–44
Loop Current, 55
Low-frequency (LF) force, 91–92
Low-frequency cyclic loads, 43

**M**
Magnetic particle inspection (MPI), 236, 245
Marine chain, 178
Mimosa, 111
Mineral Management Service, 221–222
Miner's rule, 117
Minimum breaking load (MBL), 177, 225
Minimum breaking strength (MBS), 89
Mobile moorings, 78
Mobile offshore drilling unit (MODU), 2,
        21–22, 89, 156–157, 175–176, 215,
        234, 265, 292, 294
    estimated $P_f$ for, 266–267
    improving MODU mooring reliability,
        267–268
    periodic surveys, 236
    versus permanent moorings, 236–237
    preset mooring system for, 228–229
    rig mooring system for, 226–227
    wind, wave, and current acting on, 42f

Model-of-the-model, 150
Model tests, 51−53, 139
  capability of model basin facilities,
    145−146
    current generation, 146
    wavemaker, 145−146
    wind generation, 145
  execution, 152−153
    data collection and processing, 152−153
    environment calibration, 152
    model preparation, 152
  hybrid test method, 150−151
    basic principle, 150
    numerical tools, 151
  limitations of, 147−148
  mooring system truncation, 148−150
    design, 149
    limitations due to truncation, 149
    mooring truncation, 148
  principle of, 143−145
    scale factor, 144−145
  types of, 140−142
    ice tank test, 142
    ocean basin model test, 140
    towing tank test, 141
    wind tunnel test, 141
Monitoring, 245−247
  devices, 249−251
    global positioning system−based
      system, 251
    inclinometer, 250−251
    load cell, 250
  methods, 247−249
    tension monitoring, 248
    vessel position monitoring, 248−249
    visual monitoring, 247−248
  mooring integrity management (MIM),
    293
  performance, 246−247
  regulatory requirements, 246
Mooring failures around the world, 256−261
Mooring integrity management (MIM),
  281−284
  assessing hazards and performing risk
    assessment, 283−284
  incident response, 284−287
    defining response actions, 285−286
    inclusion of procedures for readiness
      check of equipment, 287
    inclusion of sparing plan, 286−287
    predefining installation procedures
      and contracting plan, 287

  life extension, 287−291
    fitness assessment of mooring
      component, 289−291
    for floating facility and its mooring
      system, 288−289
  management of mooring performance,
    282−283
  ways to improve mooring integrity,
    291−295
    equipment with monitoring system, 293
    improving codes and standards, 294−295
    perform rigorous inspection and
      maintenance, 291−293
    sharing lessons learned, 294
Mooring line failure
  identifying and verifying, 285
Mooring Rapid Response Plan (MRRP),
  284−285
Mooring repair, 286
Mooring Risk Management Plan, 284−285
Mooring software, 110−111
  Ariane by Bureau Veritas, 111
  DeepC/SESAM by DNV GL, 110−111
  OrcaFlex by Orcina Ltd., 110
Mooring truncation, 148
Morison model, 52−53, 54*f*, 93
Morison's load formula, 51
"Motion-free" floating drilling platform,
  2−3
Motion transfer functions, 53, 140
Movable tensioning system, 204−205
Movable winches, 204−206
Movable windlass, 204−206

**N**

Navier−Stokes equations, 87
Nondestructive examination (NDE)
  techniques, 240
Nonreal-time monitoring system, 246−247
Numerical computations, 46, 51−53, 56
Numerical models, 85−86, 116−117, 150
Numerical simulation, 53, 58−59
Nylon (polyamide) rope, 187−188

**O**

Ocean basin model test, 140, 140*f*
Ocean currents, 55, 58
Ocean Driller, 2−3
Offshore drilling operations, 7−8
Offshore installation manager (OIM), 276

Offshore mooring, history of, 2–7
　floating drilling, 2–4
　floating production, 5
　industry standards, 6–7
　technologies, 5–6
Offshore mooring chain, 176–179
Offshore moorings, 2, 19, 190, 255–256
Offshore rig quality (ORQ) chain,
　177–178
Oil Companies International Marine Forum
　(OCIMF), 34–35, 47, 188
OMNI-Max anchor, 171–173, 172f
On-vessel equipment, 199, 261, 281, 287–288
　chain jack, 203–204
　chain windlass, 204–206
　　movable windlass, 204–206
　in-line tensioner, 209–212
　tensioning systems, 199–203
　　chain locker, 202–203
　　fairlead and stopper, 201
　　hydraulic or electric power unit,
　　　201–202
　wire winch, 206–209
　　drum winch, 206
　　linear winch, 209
　　traction winch, 207–209
Open-end socket, on wire rope, 181f
Optimal model scale, 144f
OrcaFlex by Orcina Ltd., 110
Out-of-plane bending (OPB), 265
　fatigued chain due to, 275
Out-of-plane bending (OPB) fatigue for
　chain, 132–135
　mechanism of, 132–134
　OPB fatigue assessment, 134–135
Overall project cost, 313

**P**

Palmgren–Miner linear damage hypothesis,
　117
Panel model, 53
Pear link, 194, 194f
PelaStar, 302–303
Permanent moorings, 6, 20, 35, 72, 95, 170,
　182, 191, 234–235, 262–264, 269,
　281, 292
　as-built survey, 235
　corrosion resistance, 182
　failures of, between 2001 and 2012, 257t
　installation, 217–225
　　Phase III—hook-up of mooring lines to
　　floating production unit, 222–225

Phase II—prelay of mooring lines on
　seabed, 219–222
Phase I—installation of pile anchors,
　217–219
versus mobile offshore drilling unit,
　236–237
periodic surveys, 235–236
probability of failure, 261–263
　estimated $P_f$ for permanent moorings,
　262–263
　failure spots for permanent moorings,
　264–265
system versus component failures, 263
Phased Array Ultrasonic Testing (PAUT),
　179f, 180
Pierson–Moskowitz spectrum, 50–51, 51f
Plastic limit analysis methods, 159
PLEM (Pipeline End Manifold), 34–35
Polyester fiber ropes, 190
Polyester mooring systems, 6, 74, 98, 185
Polyester rope, 72–73, 95, 183–187,
　219–221, 228–229
　constructions, 185–187, 186f
　first use of polyester mooring in
　　deepwater, 185
　polyester stretch, 187
　stiffness, modeling of, 97–100
　　static–dynamic model, 99–100
　　upper–lower bound model, 98–99
Position measurement system, 37
Position monitoring, 248–249
Power steering system, 37
Production semisubmersibles, 177, 212
Propeller system, 37

**Q**

Quadratic transfer functions (QTFs), 61, 110
Quasistatic or dynamic analyses,
　101–102
Quasistatic stiffness, 99–100

**R**

Radiation/diffraction programs, 93
Radiation–diffraction theory, 52
Rain-flow counting method, 122–123,
　125–126
Raised external turret mooring system, 27
Rapid response team, 285
Real ocean waves, 49–50
Real-time monitoring system, 246–247
Recurrence interval, 42

Reference breaking strength (RBS), 80, 117
Regular waves, 49, 145–146
  superposition of, 49–50, 50f
Reliability of mooring, 255
  failure mechanisms, wide variety of,
    273–276
    chain damaged from handling, 276
    chain with severe corrosion, 274
    deficient chain from manufacturing, 274
    fatigued chain due to out-of-plane
      bending, 275
    knotted chain due to twist, 275–276
    operation issues, 276
  failure spots for temporary moorings,
    269–270
  mooring components, 270–273
    percentage distribution of chain failures
      by cause, 272–273
    percentage distribution of mooring
      failures by component type, 270–272
  mooring failures around the world,
    256–261
  probability of failure for permanent
    moorings, 261–263
    estimated $P_f$, 262–263
    failure spots, 264–265
    system versus component failures, 263
  probability of failure for temporary
    moorings, 265–268
    estimated $P_f$ for MODU moorings,
      266–267
    improving MODU mooring reliability,
      267–268
Remotely operated vehicle (ROV), 160,
  192–193, 217, 235, 237–238, 265,
  285–286
Renewable energy, 299
Response actions, defining, 285–286
  condition assessment, 285–286
  emergency response, 285
  mooring repair, 286
Response amplitude operators (RAOs), 93
Response-based analysis (RBA), 109–110
Return period, 42–43, 45–46, 78
Reynolds number, 95, 147
Ridged ice, 58
Riflex, 110–111
RIM (Riser Integrity Management), 281
Risers
  modeling, 96
  umbilical systems, 21
Riser turret mooring (RTM) system, 29, 30f

Risk assessment, performing, 283–284
Root Cause Analysis (RCA), 285
Rotary winches, 201

## S

Scaled model tests, 51–52
SCORCH, 291
Sea-keeping tests, 140
Semiempirical methods, 159–160
Semisubmersible, 1, 4–5, 11, 44, 60
Semisubmersible hulls, 9
Semisubmersible type, 301–302
SESAM, 110
Shallow and deep waters
  anchor types for, 156f
Shallow-water moorings, 72–73, 93
Sheathed spiral strand wire rope, 180f, 183
Shell Oil, 2–3
Shielding coefficients, 47
Ship anchoring chain, 178
Ship-shaped floating vessels, 22
Significant wave height, 49–51, 60, 127, 152
Single-point mooring (SPM) system, 21,
  23–32
  catenary anchor leg mooring system,
    34–35
  disconnectable turret mooring system,
    28–32
  external turret mooring system, 26–28
  internal turret mooring system, 24–26
  tower yoke mooring system, 32–33
Single-point mooring system, 14, 23–32, 60,
  67, 103–105
Site investigation, 156–158, 215–216
  geophysical survey, 216
  geotechnical survey, 216
Six-strand ropes, 181–182
Six-strand versus spiral strand, 181–182
Small amplitude wave theory, 49
S–N curves, 117
  for chain and wire ropes, 119–120
Software programs, 47, 53
Soil strength, 158
Solitary wave theory, 49
Spar, 1, 5, 12–13
Spar-buoy type, 301
SPARs, 44, 177
Spectral domain calculations, 47
Spread-moored FPSO systems, 22–23, 60
Spread mooring system, 6–7, 21–23, 21f
Spread versus single-point moorings, 21

Squalls, 45–46
Static–dynamic model, 99–100, 99*f*
Static offset tests, 140, 151
Static stiffness, 97–100
Station-keeping tests, 140
Steady loads, 43
Steel buoys, 195
Stevmanta and Dennla, 166, 166*f*
Stevmanta VLA, 167
STL (submerged turret loading), 31–32
Stokes waves, 49
Storm surge current, 55–56
"Storm" legs, 266
STP (submerged turret production) system, 31–32, 32*f*
Stress concentration factors (SCF), 135
Stud and studless chain links, 176*f*
Studlink chain, 177, 187
Submerged external turret mooring system, 27–28, 27*f*
Subsea mooring connector (SMC) tools, 192–193, 193*f*
Subsurface buoy, 195
Suction anchors, 161
Suction embedded plate anchors (SEPLAs), 168–169, 169*f*
    advantages and limitations of, 168–169
    installation, 169
Suction piles, 6, 158–161, 159*f*
    holding capacity of, 159–160
    installation, 160–161
Surface buoy, 194*f*, 195–196
Surface currents, 45–46, 55
Sustained wind velocity profile, 45
Swell, 48
Swivel system, 24, 195
Symmetric narrow-band Gaussian spectrum, 48
Syntactic foam, 195
Synthetic fiber ropes, 306
Synthetic ropes, 187–191, 288, 306
System modeling, 93–96
    analysis procedure, 96
    modeling of environments and seabed, 96
    modeling of floaters, 93
    modeling of mooring lines, 94–95
    modeling of risers, 96
System versus component failures, 263

**T**

Tandem offloading arrangement, 14
Tanker hawser equipment, 34

Taut leg mooring systems, 20, 20*f*, 68, 228, 306–307
Technical papers, 46, 284, 294
Temporary catenary mooring systems, 159–160
Temporary mooring, 20, 188, 191, 225, 281
    deployment and retrieval, 225–229
    failure spots, 269–270
    versus permanent moorings, 20
Tender-assisted drilling (TAD), 7–10
Tensioning systems, 199–203, 313
    chain locker, 202–203
    fairlead and stopper, 201
    hydraulic or electric power unit, 201–202
    types, 200
Tension leg platform (TLP), 1, 5, 11, 155–156, 300
    TLP-based wind turbine floaters, 303*f*
    type, 302–303
Tension monitoring, 248
Tension–tension (TT) fatigue, 134
Theoretical background, 85–93
    governing equations of mooring line, 86–87
    mooring line dynamics, 91
    mooring line stiffness, 89–90
    mooring system, 91–93
    static solution, 87–89
Thermohaline circulation, 55
Three-dimensional imaging, 240–241
Thruster-assisted mooring system, 38, 81
"Thrusters", 6, 38
Tidal currents, 55–56
Time domain
    calculations, 47
    fatigue analysis, 125–126
    mooring analysis, 311–312
    strength analysis, 105–106
        analysis procedure, 106
        time-domain approach, 105
T–N curves, 117
    for chain, connectors and wire ropes, 118–119
    for polyester ropes, 120–121
    versus S–N curves, 121–122
Torpedo anchor, 170, 171*f*
Torpedo pile, 156
Total wind force, 47
Tower yoke mooring system, 32–33, 33*f*
Towing tank test, 141
Towing test, in progress, 142*f*
Traction winch, 207–209, 208*f*

Transverse VIM, 128
Trident link, 194
Truncated mooring system, 148
Truss spar, 12−13, 13*f*
Turret moored floating production storage and
    offloading, 15*f*
Turret-moored FPSO, 22−24
Turret-moored ship-shaped floaters, 38

**U**
UHC (Ultimate Holding Capacity), 165
Ultradeep waters, considerations for moorings
    in, 190−191
Uncoupled analysis, 107−109, 107*f*
Upper−lower bound model, 98−99

**V**
Vane tests, 158
Vertically loaded anchors (VLAs), 166−168,
    166*f*, 228−229
  holding capacity, 167−168
  installation, 168
  for permanent and temporary moorings,
    166−167
Vessel offset, 8*f*, 283
Vessel position monitoring, 248−249
Visual monitoring, 247−248
Vortex-induced motion (VIMs), 54, 56−57,
    57*f*, 128−131
  mechanism of, 128−129
  VIM fatigue assessment, 79, 129−131
Vortex-induced vibration testing, 141

**W**
WADAM, 52
WAMIT, 52
Wave diffraction analysis, 53
Wave frequency (WF) motion, 91
Wave frequency cyclic loads, 43
Wave generation capability, 145−146
Wave height, 49, 60, 127
Wave-induced forces and motions, 51−54
Wave length, 49
Wavemaker, 145−146, 146*f*
Wave period, 49
Wave propagation direction, 49
Waves and swells, description of, 48−51
Wave theories, 49
Western Explorer, 2, 3*f*
WindFloat FOWT, 300−302
Wind generation, 145, 145*f*
Wind load, 44−47
  description of winds, 44−46
  wind-induced forces, 46−47
Wind tunnel test, 141, 141*f*
Wind waves, 48
Wire rope, 72, 180−183
  constructions, 181*f*
  corrosion protection, 182−183
  inspection of, 243−244
  six-strand versus spiral strand, 181−182
  termination with sockets, 183
Wire strand core, 181
Wire winch, 200, 206−209
  drum winch, 206
  linear winch, 209
  traction winch, 207−209